CAMBRIDGE LIBRARY COLLECTION

Books of enduring scholarly value

Technology

The focus of this series is engineering, broadly construed. It covers technological innovation from a range of periods and cultures, but centres on the technological achievements of the industrial era in the West, particularly in the nineteenth century, as understood by their contemporaries. Infrastructure is one major focus, covering the building of railways and canals, bridges and tunnels, land drainage, the laying of submarine cables, and the construction of docks and lighthouses. Other key topics include developments in industrial and manufacturing fields such as mining technology, the production of iron and steel, the use of steam power, and chemical processes such as photography and textile dyes.

A Treatise on Roads

The politician Sir Henry Parnell (1776–1842) was instrumental in drafting legislation to improve the important road linking London with Holyhead in Anglesey, a major port for communication with Dublin. He was aided by the pioneering civil engineer Thomas Telford, and in 1833 Parnell published the first edition of this thorough work on road construction and maintenance. Reissued here is the second edition of 1838. Drawing on his experiences with Telford, who called the work 'the most valuable Treatise which has appeared in England' on the subject, Parnell outlines not only the rules governing the planning of a new road, but also addresses the practical aspects of building and repairing roads, noting the various tools and materials needed. Parnell, later Baron Congleton, also highlights the connection between road construction and national development, and includes a number of appendices relating to contemporary legislation on the subject of roads.

Cambridge University Press has long been a pioneer in the reissuing of out-of-print titles from its own backlist, producing digital reprints of books that are still sought after by scholars and students but could not be reprinted economically using traditional technology. The Cambridge Library Collection extends this activity to a wider range of books which are still of importance to researchers and professionals, either for the source material they contain, or as landmarks in the history of their academic discipline.

Drawing from the world-renowned collections in the Cambridge University Library and other partner libraries, and guided by the advice of experts in each subject area, Cambridge University Press is using state-of-the-art scanning machines in its own Printing House to capture the content of each book selected for inclusion. The files are processed to give a consistently clear, crisp image, and the books finished to the high quality standard for which the Press is recognised around the world. The latest print-on-demand technology ensures that the books will remain available indefinitely, and that orders for single or multiple copies can quickly be supplied.

The Cambridge Library Collection brings back to life books of enduring scholarly value (including out-of-copyright works originally issued by other publishers) across a wide range of disciplines in the humanities and social sciences and in science and technology.

A Treatise on Roads

Wherein the Principles on which Roads Should be Made are Explained and Illustrated, by the Plans, Specifications, and Contracts Made Use of by Thomas Telford, Esq., on the Holyhead Road

Henry Parnell

CAMBRIDGE
UNIVERSITY PRESS

University Printing House, Cambridge, CB2 8BS, United Kingdom

Cambridge University Press is part of the University of Cambridge.
It furthers the University's mission by disseminating knowledge in the pursuit of education, learning and research at the highest international levels of excellence.

www.cambridge.org
Information on this title: www.cambridge.org/9781108071741

This edition first published 1838
This digitally printed version 2014

ISBN 978-1-108-07174-1 Paperback

A TREATISE ON ROADS;

WHEREIN

THE PRINCIPLES ON WHICH ROADS SHOULD BE MADE

ARE EXPLAINED AND ILLUSTRATED,

BY

THE PLANS, SPECIFICATIONS, AND CONTRACTS

MADE USE OF BY

THOMAS TELFORD, ESQ.

ON THE HOLYHEAD ROAD.

Second Edition.

BY

THE RIGHT HONOURABLE

SIR HENRY PARNELL, BART.

HONORARY MEMBER OF THE INSTITUTION OF CIVIL ENGINEERS, LONDON.

LONDON:

PRINTED FOR

LONGMAN, ORME, BROWN, GREEN, AND LONGMANS,

PATERNOSTER-ROW.

1838.

London:
Printed by A. Spottiswoode,
New-Street-Square.

PREFACE

TO THE SECOND EDITION.

SINCE the first edition of this Treatise on Roads was published, a work has appeared, entitled "The Life of Thomas Telford, Esq., written by himself," which contains the following passages relative to this Treatise, and to the circumstances which afforded the author that particular kind of knowledge which enabled him to write it:—

"It has already been stated, that, on Lord Oriel's retiring from public life, Sir Henry Parnell, member of parliament for Queen's County in Ireland, and since for Dundee in Scotland, was not only the principal instrument in carrying the Holyhead Road Bills through parliament, but has ever since continued to be the most efficient of the commissioners. Fully impressed with the importance of rendering the communication between London and Dublin perfect, he has, during the last twenty years, applied himself to this object, for effecting which both talents and management have been required, as well as perseverance.

"1st, He had to convince government of the advantages to be derived from the scheme, and induce them to furnish the means of defraying the expense; 2d, to procure the consent of all the numerous, and in some instances turbulent, bodies of local trustees, upon an extensive line of road; and, 3d, to arrange the sea communication between Holyhead and Dublin; for which purpose the harbour of Holyhead was improved, in a manner which has rendered it serviceable as a harbour of refuge, far beyond the immediate purpose of protection of the packets; and a harbour has also been made at Howth, northward of the city of Dublin. All this he has effectually accomplished, and by extending his services beyond the usual duties of a parliamentary commissioner, and therein devoting much of his time to the personal inspection of practical operations, he has acquired so perfect a knowledge of road-making in all its branches, as has enabled him to produce the most valuable Treatise which has appeared in England, on the history, principles, and practice of that species of national improvement."

TABLE OF CONTENTS.

CHAP. III.

FORMING A ROAD, P. 82.

CHAP. IV.

DRAINAGE, P. 94.

CHAP. V.

DIFFERENT KINDS OF ROADS, AND MODES OF CONSTRUCTING THEM, P. 100.

CHAP. VI.

FENCES, P. 172.

CHAP. VII.

ROAD MASONRY, P. 176.

CHAP. VIII.

MANAGEMENT OF ROAD WORKS, P. 206.

CHAP. IX.

IMPROVING OLD ROADS, P. 234.

CHAP. X.

REPAIRING ROADS, P. 243.

CHAP. XI.

ROAD INSTRUMENTS AND TOOLS, P. 253.

CHAP. XII.

ROAD LEGISLATION, P. 263.

CHAP. XIII.

CARRIAGES, P. 295.

APPENDIX.

NOTES.

A

TREATISE ON ROADS.

INTRODUCTION.

When society has attained a high degree of industry and wealth, so many persons and goods are set in motion for the purpose of administering to its business and its luxuries, that the construction of public roads in such a manner as to admit of safe and rapid transport, and of the reduction of the cost of carriage to the lowest possible amount, becomes matter of the greatest importance.

To explain how these objects may be best secured is the purpose of the following pages.

The measures necessary for affording the means of travelling with rapidity and safety, and of transporting goods at low rates of carriage, form an essential part of the domestic arrangements of every people. Roads are, in point of fact, necessary ingredients in the first change that every rude country must undergo in emerging from poverty and barbarism. It is, therefore, one of the most important duties of government to enact such laws, and provide such means, as are requisite for the structure and maintenance of well-made roads throughout the territory under its authority.

M. Storch most correctly says, that, " after giving protection to property and person, a government can bestow on a nation no greater benefit than the improvement of its harbours, canals, and roads." *

Speaking of roads, the Abbé Reynal justly remarks, " Let us travel over all the countries of the earth, and whenever we shall find no facility of travelling from a city to a town, or from a village to a hamlet, we may pronounce the people to be barbarians."

It has been well said by a writer in the first volume of the Communications to the Board of Agriculture, that " the conveniencies and beneficial consequences which result from a free and easy communication between different parts of a country are so various, the advantages of them so generally and so extensively felt by every description of individuals from the highest to the lowest, that no labour or expense should be spared in providing them. Roads, canals, and navigable rivers, may be justly considered as the veins and arteries through which all improvements flow. How many places in almost every country might be rendered

* Cours d'Economie Politique, vol. i. p. 188.

doubly valuable, if access to them were practicable and easy!"

Adam Smith says, "Good roads, canals, and navigable rivers, by diminishing the expense of carriage, put the remote parts of a country nearly on a level with those in the neighbourhood of a town; they are, upon that account, the greatest of all improvements."

To establish perfect roads throughout a country is an object of no small importance as regards public economy. In proportion as roads are level and hard, will there be a saving of horse labour, a cheaper description of horse may be employed, less food will be consumed, and fewer servants wanted. The expense of travelling, and the charges for the carriage of goods, will be lower. A saving to the public, amounting in the aggregate to a considerable sum, will thus annually take place, to be applied either to the accumulation of national capital, or to some other purpose.

It will be useful, previously to showing what is necessary to be done in order to secure good roads in this country, to mention the conduct of other nations in this branch of domestic economy.

A description of this kind may serve to give a better tone to the ideas of those country gentlemen, who are trustees of the public in the management of its roads, and may encourage them to form a more enlarged, and more correct conception of their duties and their responsibility.

The following quotations are taken from the French Encyclopædia, under the head of *Chemin.*

The very interesting information they contain will be a sufficient apology for their length :—

"The police of roads does not begin to show itself as worthy of consideration until the prosperous times of Greece. The senate of Athens watched over them. The Lacedæmonians, Thebans, and other states, confided them to the care of the most eminent men. It does not, however, appear that this display of management produced any considerable effect in Greece. It was reserved for a commercial people to observe the benefits of facility of travelling and transporting goods; hence it is that the invention of paved roads is given to the Carthaginians.

"The Romans did not neglect the example of the Carthaginians, and that part of their labours is not the least glorious to this people. The first road they made was the *Via Appia*, the second the *Via Aurelia*, the third the *Via Flaminia*. The public and the senate held the roads in such estimation, and took so great an interest in them, that under Julius Cæsar the principal cities of Italy all communicated with Rome by paved roads.

"Their roads from that period began to be extended into the provinces.

"During the last African war, the Romans made a road with rectangular broken stones (de cailloux taillés en quarré), from Spain through Gaul to the Alps. Domitius Œnoberbus paved the *Via Domitia*, which led to Savoy, Dauphiny and Provence. The Romans made in Germania another *Via Domitia*.

“ Augustus, when emperor, paid more attention to the great roads than he had done during his consulate. He conducted roads into the Alps; his plan was to continue them to the eastern and western extremities of Europe. He gave orders for making an infinite number in Spain; he enlarged and extended the *Via Medina* to Gades. At the same time, and through the same mountains, there were opened two roads to Lyons; one of them traversed the Tarentaise, and the other was made in the Alphenin.

“ Agrippa seconded Augustus ably in this part of his government. It was at Lyons he began the extension of roads throughout all Gaul.

“ There are four of them particularly remarkable for their length, and the difficulty of the country through which they passed. One traversed the mountains of Auvergne, and penetrated to the bottom of Aquitaine. Another was extended to the Rhine at the mouth of the Meuse, and followed the course of the river, to the German Ocean; the third crossed Burgundy, Champagne, and Picardy, and ended at Boulogne-sur-Mer; the fourth extended along the Rhone, entered the bottom of Languedoc, and terminated at Marseilles. From these principal roads there were an infinity of branch roads, namely, to Tréves, Strasburg, Belgrade, &c.

“ There were also great roads from the eastern provinces of Europe to Constantinople, and into Croatia, Hungary, Macedonia, and to the mouth of the Danube at Torres.

"The seas were able to cut across the roads undertaken by the Romans, but not to stop them. Witness Sicily, Corsica, Sardinia, England, Asia and Africa, the roads of which countries communicated with the roads of Europe by the nearest ports. What labours! when we embrace in one point of view the extent and the difficulties which opposed themselves; the forests opened, the mountains cut through, the hills lowered, the valleys filled up, the marshes drained, and the bridges which were built!" *

The following description of the manner in which the Roman roads were made, is taken from the same work: —

"Les grands chemins étoient construits selon la diversité des lieux: ici ils s'avançoient de niveau avec les terres; là ils s'enfonçoient dans les vallons; ailleurs ils s'élevoient à une grande hauteur; partout on les commençoit par deux sillons tracés au cordeau; ces parallèles fixoient la largeur du chemin; on creusoit l'intervalle de ces parallèles; c'étoit dans cette profondeur qu'on étendoit les couches des matériaux du chemin. C'étoit d'abord un ciment de chaux et de sable de l'épaisseur d'un pouce; sur ce ciment, pour première couche, des pierres larges et plates, de dix pouces de hauteur, assises les unes sur les autres, et liées par un mortier des plus durs; pour seconde couche, une épaisseur de huit pouces de petites pierres rondes, plus

* French Encyclopædia, article Chemin, folio edit. vol. iii. p. 276.

tendres que le caillou, avec des tuiles, des moillons, des platras, et autres décombres d'édifice, le tout battu dans un ciment d'alliage; pour la trosième couche, un piéd d'épaisseur d'un ciment fait d'une terre grasse, mêlée avec de la chaux; ces matières intérieures formoient depuis trois piéds jusqu'à trois pièds et demi d'épaisseur. La surface étoit de gravois lié par un ciment mêlé de chaux; et cette croûte a pu résister jusqu'à présent en plusieurs endroits de l'Europe."

"The Roman roads," says Mr. Tredgold, "ran nearly in direct lines; natural obstructions were removed or overcome by the effort of labour or art, whether they consisted of marshes, lakes, rivers, or mountains. In flat districts, the middle part of the road was raised into a terrace.

"In mountainous districts, the roads were alternately cut through mountains and raised above the valleys, so as to preserve either a level line or a uniform inclination. They founded the road on piles where the ground was not solid, and raised it by strong side walls, or by arches and piers where it was necessary to gain elevation. The paved part of the great military roads was sixteen Roman feet wide, with two side ways, each eight feet wide, separated from the middle way by two raised paths of two feet each." *

The funds for making roads were so well secured and so considerable, that the Romans were not

* See Tredgold on Railways, p. 6.

satisfied to make them convenient and durable, but they also embellished them.

"They had columns placed from mile to mile to mark the distance of one place from another; blocks of stone for foot travellers to rest upon, and to assist horsemen to mount their horses; and also temples, triumphal arches, mausoleums, and military stations.

"Such was the solid construction of the Roman highways, that their firmness has not entirely yielded to the effect of fifteen centuries.

"They united the subjects of the most distant provinces by an easy and familiar intercourse; but their primary object was to facilitate the marches of their legions." *

After enumerating all the cities in the different parts of the empire, Mr. Gibbon says, "All these cities were connected with each other and with the capital by the public highways, which, issuing from the forum of Rome, traversed Italy, pervaded the provinces, and were terminated only by the frontiers of the empire.

"If we carefully trace the distance from the wall of Antoninus (in Britain) to Rome, and from thence to Jerusalem, it will be found that the great chain of communication from the north-west to the south-east part of the empire was drawn out to a length of 4080 Roman miles, or 3740 English miles. The public roads were accurately divided by milestones,

* See Bergier, Histoire des grands Chemins de l'Empire Romain, liv. ii. cap. 1. p. 28.

and ran in a direct line from one city to another, with very little respect for the obstacles either of nature or private property: mountains were passed, and bold arches thrown over the broadest and most rapid streams. The middle part of the road was raised into a terrace, which commanded the adjacent country, and consisted of several strata of sand, gravel, and cement, and was paved with large stones, which in some places near the capital were of granite."

The following are Mr. Pinkerton's observations on the Roman roads: —

"One of the grand causes of the civilisation introduced by that ruling people (the Romans) into the conquered states, was the highways, which form, indeed, the first germ of national industry, and without which neither commerce nor society can make any considerable progress. Conscious of this truth, the Romans seem to have paid particular attention to the construction of roads in the distant provinces; and those of England, which may still be traced in various ramifications, present a lasting monument of the justice of their conceptions, the extent of their views, and the utility of their power. A grand trunk, as it may be called, passed from the south to the north, and another to the west, with branches in almost every direction that general convenience and expedition could require. What is called Watling Street led from Richborough, in Kent, the ancient Rutupiæ, north-east, through London, to Chester. The

Ermine Street passed from London to Lincoln, thence to Carlile, and into Scotland.

"The Foss Way is supposed to have led from Bath and the Western regions, north-east, till it joined the Ermine Street. The last celebrated road was Ikeneld, or Ikneld, supposed to have extended from near Norwich, southward, into Dorsetshire."*

Mr. Eustace says, in his Classical Tour, "Thus the civilised world owes to the Romans the first establishment and example of a commodious intercourse; one of the greatest aids of commerce and means of improvement that society can enjoy."

He further says, "The barbarians who overturned the Roman power were for many ages so incredibly stupid as to undervalue this blessing, and almost always neglected, and sometimes wantonly destroyed the roads that intersected the provinces which they had invaded. To this day the different governments of Germany (except Austria), Spain, Portugal, Sicily, and Greece, are still so immersed in barbarism, as to leave the traveller to work his way through their respective territories, with infinite fatigue and difficulty, by tracks and paths oftentimes almost impracticable."†

Among modern nations, France is one of the most distinguished for her early attention to establishing numerous roads.

* Pinkerton's Geography, vol. i. p. 20.

† Vol. iii. p. 178.

The following account of her roads is taken from Peuchet's Statistical Account of France:—"The origin of our principal roads is generally attributed to Philip Augustus; it was under his reign, and by his orders, that the city of Paris began to be paved.

"Sully took great interest in the improvement of the roads. He first introduced the practice of planting trees on the sides of them, and established regular funds for their repair. Colbert neglected nothing to advance the extension of roads throughout France; and M. Desmartis, who succeeded him, caused the road from Paris to Orleans to be made. He was the founder of the corps of engineers, appointed to superintend the works belonging to the roads. Under the administration of the Duke de Noailles, the roads were improved and carried through the provinces. In 1726 the department of the *Ponts et Chaussées* fell into great disorder, and was in want of sufficient funds, but the Director-general, the brother of the celebrated Cardinal Dubois, recommenced the repairs of them, and continued them with great regularity.

"Under the administration of M. de Trudaine, in 1787, a number of new roads were made. He established the *Ecole des Ponts et Chaussées,* under M. Perronet, as chief engineer, and at his death he left to this school his manuscripts and library. This school is under the minister of the Home Department; the scholars are fifty in number; these are selected from the Polytechnic School, and receive an allowance of seventy-five francs a month.

"The roads of France were divided at this time into four classes, according to their importance, and the breadth that is given to them. The first class comprised the great roads which traverse the whole of France, from Paris to the principal cities and the ports; the second class, the roads between the provinces and principal cities; the third class, the roads between the principal towns in the same province and the neighbouring provinces; and the fourth class, the roads between small towns and villages.

"By an order of council of the 6th of February, 1776, the breadth of the first class was fixed at forty-two feet (French) between the fences; of the second at thirty-six feet; of the third at thirty feet; and of the fourth at twenty-four feet.

"The roads have since been divided into three classes, not according to their breadth, but their direction."*

All the principal roads of France are under the management of Government. The department of the *Ponts et Chaussées* has the care of them. In the year 1836, the sum of £896,000 was granted by the chambers for maintaining them.

Notwithstanding, however, the attention which has been paid to the roads in France, their actual state as to number, extent, and condition, shows that the system of management is extremely imperfect.

With the exception of those parts of the main roads leading from Paris which are paved, they are

* Peuchet, p. 458.

weak and rutted. In those districts where they are repaired with gravel, they are almost impassable in winter; the diligences with six horses with difficulty travel four miles an hour. In other districts, where the materials are harder, a road is seldom to be seen with a smooth surface and of sufficient strength. Very extensive tracts of the kingdom are wholly without regularly formed roads; and, therefore, however valuable the efforts of the statesmen of France may have been in bringing the business of road-making to the point at which it has arrived, there is still wanting some new plan of legislation, by which good roads may be made, not only from one town to another, but into and through every commune in France.

In Spain the *caminos reales*, or king's highways, are not numerous, nor are they kept in good repair. Taking Madrid as a point of departure, there are two good roads to Burgos; one passing through Valladolid, and the other through Aranda de Duero. From Burgos, the road is continued by Vittoria and Irun to France. Both these roads are in tolerable repair. From Valladolid a good road has been made by Valentia and Reynosa to Santander. There are two good roads to Bilboa; one by Miranda, the other by Vittoria.

To the northward, there is a *camino reale* through Gallicia to Corunna and Ferrol; but in such want of repair, as to be impassable in numerous places for loaded carriages; attempts are, however, now making to improve it. In Catalonia the roads are comparatively numerous and good.

The road from Saragosa to Barcelona has lately been put in repair, and a diligence was established upon it in the beginning of the year 1831.

The other roads which are traced upon maps of Spain may be divided into three classes:—

1st, Roads which have originally been made and covered with road metal; 2dly, Roads across the plains and through the valleys, formed by the tracks of the country carts, and which have only, in a few places, been artificially constructed; and, 3dly, the mule roads or paths, worn by the feet of the mules travelling over the mountains during a long series of years.

The revenue applicable to the construction and repair of the roads is derived, 1st, from tollgates; and 2dly, from local taxes. Upon all the practicable roads tolls are established at intervals of ten or twelve English miles.*

The following remark on the roads of Spain is taken from the Edinburgh Review for July, 1832:—"Another check upon agriculture is, that with the exception of some few high roads, which are sufficiently insecure, there exists scarce a waggon or cart track throughout Spain. All means of transport are therefore dear; and in Salamanca, it has been known, after a succession of abundant harvests, that the wheat has actually been left to rot, because it would not repay the cost of carriage. About 90,000*l.* is the average annual expenditure upon the roads in Spain."†

* See Foreign Quarterly Review, vol. i. p. 82.

† Vol. lxv. p. 448.

In the most populous districts of the German and Russian dominions, the chaussée, or paved road, similar to that of France, is common; but over a great part of these countries the roads are little more than formed, being almost without any prepared surface. The roads in Holland are generally carried in undeviating straight lines along that low and flat country, between a double row of trees, with a ditch on each side. The Dutch are at great pains in preparing a firm foundation for their roads; which are then built with their bricks, called *clinkers*, laid in lime; their longest direction being across the road. The Swedes have long had the character of being excellent road engineers. Good rock is very generally met with in Sweden, and they spare no pains in breaking it small; their roads are spacious and smooth.

Where the country has been opened in Russia the roads are formed on scientific principles, but there are few of them. In the United States of America the roads have latterly been much improved; the principal lines are similar to the generality of those in England. Italy still preserves its celebrity for interior communication.*

Before the peace of 1814, there was but one great road throughout the kingdom of Prussia, namely, that between Berlin and Magdeburgh, a distance of thirty leagues; the rest were scarcely practicable, and kept in a most detestable state.

* The foregoing description of foreign roads is taken from Mr. Stevenson's article on roads in the Edinburgh Encyclopædia.

There are now a number of great roads communicating between the capital and various parts of the kingdom, kept in the best order; most of them at the expense of the government, and a few defrayed by the local authorities. In the towns and villages through which these roads pass, the pavement is generally in a very bad state; the expense being paid by the municipal authorities, who are very independent, and only repair them when it suits their convenience.*

The little attention that was paid in former times to the roads of England is made evident by a proclamation of Charles the First, issued in 1629, confirming one of his father's, issued in the twentieth year of his reign, for the preservation of the roads of England, which commands "that no carrier or other person whatsoever shall travel with any waine, cart, or carriage with more than two wheels, nor with above the weight of twenty hundred, nor shall draw any waine, cart, or carriage, with more than five horses at once."†

The first attempt to put the roads of England into order occurred when the turnpike system was introduced. The ancient method employed to mend roads, until after the restoration of King Charles II., was by a pound rate on the landholders in the respective counties; and by the supply of carts and horses by parishes, for a limited number of days. But when, after the last named

* Vol. xii. p. 511. of Foreign Quarterly Review on *Notes et Réflexions sur la Prusse en* 1833, par le Marquis de Chambray.

† Anderson's Commerce, vol. xix. p. 130.

period, commerce so generally increased, and in consequence thereof wheel-carriages and pack-horses were extremely multiplied, the first turnpike road was established by law (the 16 Charles II. cap. 1. anno 1653), for taking toll of all but foot passengers on the northern road through Hertfordshire, Cambridgeshire, and Huntingdonshire; which road was then become very bad, by means of the great loads of barley and malt, &c. brought weekly to Ware in waggons and carts, and from thence conveyed by water to London. "These roads," says the act, "were become so ruinous and almost impassable, that the ordinary course appointed by all former laws and statutes of this realm is not sufficient for the effectual repairing of the same, neither are the inhabitants, through which the said roads lie, of sufficient ability to repair the same," &c. &c. Wherefore three tollgates were erected, one for each of these three counties, viz. at Wadesmill, Caxton, and Stilton.*

It was not, however, till after the peace of 1748 that any thing like a great exertion was made to change the public highways from the wretched state in which they had always been.

The following description of the roads is taken from Mr. M'Culloch's Dictionary of Commerce:—

"It is not easy for those accustomed to travel along the smooth and level roads by which every part of the country is now intersected to form an

* Anderson's Commerce, vol. v. p. 44.

accurate idea of the difficulties the traveller had to encounter a century ago.

"Roads were then hardly formed, and in summer not unfrequently consisted of the bottoms of rivulets. Down to the middle of the last century most of the goods conveyed from place to place in Scotland, at least where the distances were not very great, were carried, not by carts or waggons, but on horseback. Oatmeal, coals, turf, and even straw and hay, were conveyed in this way. At this period, and for long previous, there were a set of single-horse traffickers (cadgers) that regularly plied between different places, supplying the inhabitants with such articles as were then most in demand, as salt, fish, poultry, eggs, earthenware, &c.; these were usually conveyed in sacks or baskets, suspended one on each side the horse. But in carrying goods between distant places it was necessary to employ a cart, as all that a horse could carry on his back was not sufficient to defray the cost of a long journey. The time that the *carriers* (for such was the name given to those that used carts) usually required to perform their journeys seems now almost incredible. The common carrier from Selkirk to Edinburgh, *thirty-eight* miles distant, required a *fortnight* for his journey between the two places, going and returning! The road was originally among the most perilous in the whole country: a considerable extent of it lay in the bottom of that district called the Gala Water, from the name of the principal stream, the channel of the water being, when not flooded, the track chosen as the most level and easiest to travel in. Even between the largest cities the means of travelling were but little superior. In 1678 an agreement was made to run a coach between Edinburgh and Glasgow, a distance of forty-one miles, which was to be drawn by *six* horses, and to perform the journey from Edinburgh to Glasgow and back again in six days. Even so late as the middle of the last century, it took a day and a half for the stage coach to travel from Edinburgh to Glasgow, a journey which is now accomplished in four and a half or five hours.

" So late as 1763 there was but one stage coach from Edinburgh to London, and it set out only once a month, taking from twelve to fourteen days to perform this journey! At present, notwithstanding the immense intercourse between the two cities, by means of steam packets, smacks, &c., six or seven coaches set out each day from the one or the other, performing the journey in from forty-five to forty-eight hours."*

Mr. Arthur Young, in his Six Months' Tour, published in 1770, gives the following account of some of the roads in the north of England:—" *To Wigan.* Turnpike.—I know not in the whole range of language terms sufficiently expressive to describe this infernal road. Let me most seriously caution all travellers who may acci-

* M'Culloch's Dictionary of Commerce, art. Roads.

dentally propose to travel this terrible country to avoid it as they would the devil, for a thousand to one they break their necks or their limbs, by overthrows or breakings down. They will here meet with ruts, which I actually measured four feet deep, and floating with mud only from a wet summer; what therefore must it be after a winter? The only mending it receives is tumbling in some loose stones, which serve no other purpose than jolting a carriage in the most intolerable manner. These are not merely opinions, but facts; for I actually passed three carts, broken down, in these eighteen miles of execrable memory. *To Warrington.* Turnpike.—This is a paved road, most infamously bad; any person would imagine the people of the country had made it with a view to immediate destruction! for the breadth is only sufficient for one carriage, consequently it is cut at once into ruts; and you may easily conceive what a break down, dislocating road ruts, cut through a pavement, must be.

" *From Dunholm to Knutsford.* Turnpike.—It is impossible to describe these infernal roads in terms adequate to their deserts. *To Newcastle.* Turnpike.— A more dreadful road cannot be imagined. I was obliged to hire two men at one place to support my chaise from overturning. Let me persuade all travellers to avoid this terrible country, which must either dislocate their bones with broken pavements, or bury them in muddy

"It is only bad management that can occasion such very miserable roads, in a country so abounding with towns, trade, and manufactures." *

Mr. Chambers says, in his estimate, "Turnpikes which we saw first introduced soon after the Restoration were erected slowly, in opposition to the prejudices of the people. The act which for a time made it felony at the beginning of the reign of George the Second to pull down a tollgate was continued as a perpetual law before the conclusion of it. Yet the great roads of England remained almost in their ancient condition, even as late as 1752 or 1754, when the traveller seldom saw a turnpike for 200 miles after leaving the vicinity of London." †

After 1760 the general spirit of improvement led to that of the roads; and in the fourteen years from that period, to 1774, no less than 452 Turnpike Acts were passed. Since that year a number of Turnpike Acts have continued to be passed, as will appear from the following table ‡:—

In eight years, from

1785 to 1792	- -	302
1792 to 1800	- -	341
1800 to 1809	- -	419

In every year since 1809 turnpike roads have been progressively established, till they have extended to nearly 23,000 miles.

* Vol. iv. pp. 430. 434. † P. 124.

‡ Encyclopædia Brit. art. England, vol. iv. p. 112. Second Supplement.

But although this turnpike system has conduced to make many new roads, and to change many old ones into what may be called good roads, in comparison with what they formerly were, this system has been carried into execution under such erroneous rules, and the persons who have been intrusted with the administration of them have uniformly been either so negligent, or so little acquainted with the business of making or repairing roads, that at this moment it may be stated, with the utmost correctness, that there is not one in England, except those recently made by some eminent civil engineers, which is not extremely defective in its most essential qualities.*

With regard to the lines of direction of the turnpike roads, they evidently have not been laid out according to any fixed principle; they are, in fact, almost in every instance, precisely the identical lines, which formed the footpaths of the aboriginal inhabitants of the country.

The following passage is taken from a pamphlet, called "The Landed Property of England:"—"Most of the old roads of the kingdom (the remains of the Roman ways excepted) owe their present lines to particular circumstances. Many

* Mr. Edgeworth says, in his *Essay on the Construction of Roads*, published in 1813, "Since this Essay was written, I have visited England, and have found, on a journey of many hundred miles, scarcely twenty of well-made road. In many parts of the country, and especially near London, the roads are in a shameful condition; and the pavement of London is utterly unworthy of a great metropolis." Introduction, p. 7.

of them were, no doubt, originally footpaths; some of them, perhaps, the tracks of the aboriginal inhabitants, and these footpaths became, as the condition of society advanced, the most convenient horsepaths. According as the lands of the kingdom were appropriated, the tortuous lines of road became fixed and unalterable, there being no other legal lines left for carriage roads, and hence the origin of the crookedness and steepness of existing roads."

As many other great defects exist in all the principal roads, it is to be hoped, that at length the attention of the public and of government will be roused, and seriously and effectually engaged in bringing about a proper remedy. These defects are, in point of fact, so numerous and so glaring, that it is quite evident that the true principles of the art of road-making have not yet been followed. The breadth of a road is seldom defined to a regular number of feet by straight and regular boundaries, such as fences, footpaths, mounds of earth, or side channels. The transverse section of the surface, when accurately ascertained by taking the levels of it, is rarely to be found of a regular convexity. The surface of all the roads, until within a few years, was every where cut into deep ruts, and even now, since more attention has been paid to them, though the surface is smoother, the bed of materials which forms it is universally so thin, that it is weak, and consequently exceedingly imperfect. Drainage is neglected; high hedges and trees are allowed to intercept the action of the sun and wind in drying the surface roads; and many roads, by constantly carrying off the mud from them for a number of years, have been sunk below the level of the adjoining fields, so that they are always wet and damp, and owing to the rapid decay of the materials which are laid upon them, can be kept in order but at a great expense.

The business of road-making in this country has been confined almost entirely to the management of individuals wholly ignorant of the scientific principles on which it depends. It has received, till very lately, little attention from the scientific world; so little, indeed, that the primary and indispensable objects of providing a dry and sound foundation for the surface materials, and of giving the surface a regular convexity, have not, till within a short time, been recognized or explained by any scientific rules whatever. Although various select committees of the House of Commons have been appointed to take into consideration the state and condition of the roads, it does not, however, appear that any system for forming them on scientific principles has been suggested by them. On the contrary, the approbation expressed in their reports of the doctrines of a modern publication as comprising a perfect system of road-making, shows that they were not qualified for this task: for nothing, in point of fact, can be more opposed to the principles of science with respect to moving bodies, such as carriages, on roads, than what is recommended in that work as the perfection of road-making.*

In point of fact, there has not been any book in the English Language that treats of the science and art, or, in other words, of those facts and those rules of civil engineering which are applicable to the construction of roads. The various publications on this subject are none of them the works of civil engineers. The tracing of new lines, cutting through hills, and forming embankments, making the breadth every where the same, and defining it

* Remarks on Road-making, by John Loudon M'Adam.

Extract from the Westminster Review, Vol. iv. *p.* 354.

"We consider this latter person's name (Mr. M'Adam) as deserving of a few remarks, for other reasons than its present popularity. The public naturally looks on him as a sort of magician, and his invention, as it is thought, as something preternatural. If his own name had not been Macadmizable into a verb, it is probable that his roads would, even yet, have been little known. He did not invent the method in question of breaking stone, because it had long been the practice of Sweden and Switzerland, and other countries, and was long known to every observing traveller.

Although what is commonly called Mr. M'Adam's system of road-making had nothing original in it, and was in all its essential points not only not according to the principles of science, but directly opposed to them, still he certainly had the merit, of no inconsiderable value, of being the first person who succeeded in persuading the trustees of turnpike roads to set seriously about the improvement of them. By teaching them how to prepare materials, and keep the surface of a road free from ruts by continually raking and scraping it, he produced a considerable change for the better in all the roads of the kingdom.

with side channels and footpaths; giving roads a regular transverse shape, so that the surface shall have an uniform convexity, and that the side boundaries shall be on the same level; and so constructing the surface as to reduce the intensity of the tractive power as much as possible, have all been wholly overlooked.

The foreign scientific traveller must be astonished to find that a nation like England, which displays such an extent of science as regards its canals, docks, bridges, and other public works, should exhibit in its roads such great imperfections. It must, in truth, be admitted that the roads of England do no credit to the wisdom of her laws respecting them, nor to the care and skill of those who have been intrusted with their management.

While, during a considerable number of years, every improvement which depended on the industrious classes made immense progress, that of roads, the management of which the laws have vested in the hands of the land proprietors, made no advancement at all until very recently.

It was only about fifteen years ago that the landed proprietors seem to have begun to comprehend the value of good roads, and to be aware that large funds, and a considerable share of science and constant attention are necessary to bring them into a perfect state.

At the present time, although the country gentlemen are somewhat more active and better informed, the degree of improvement which they have introduced is little more than the palliative of a great evil, and goes but a short way towards securing that perfection which ought to be universally introduced.

While, however, England has been content to suffer her roads to be in so defective a state, Scotland and Ireland have acted far otherwise, and for a long time have had the benefit of a more improved system.

Lord Daer, eldest son of the Earl of Selkirk, about the year 1790, introduced into Scotland, and more especially through Galloway, the practice of laying out roads with the spirit level. The road from Dumfries to Castle Douglas was traced out by him, so as to have no greater inclination than 1 in 40, although passing through a very hilly country. Mr. Abercromby pursued, as a regular profession, the business of making roads. He laid out the road from Kinross to Perth, and by following the valleys, obtained an excellent line, instead of one passing over a successive range of very steep hills. He also laid out the road from Perth to Dunkeld. In all cases he acted on the principle of never making a road to ascend a single foot unless absolutely unavoidable; and this he accomplished by taking advantage of the valleys of the country, and skilfully cutting through high banks and filling hollows.

Mr. Abercromby made all his roads with stones broken very small. This practice had long existed in Scotland, and is recommended by the old writers on roads in France.

Since his time, many others have followed the same rules; so that Scotland has been making great progress for a number of years in establishing excellent roads.

In Ireland the abolition of the system of statute labour in 1763, and the placing of the business of making roads under the grand juries, immediately led to a great improvement; for, notwithstanding the abuses which have attended the exercise of the powers of grand juries in money matters, the general result has been to establish good roads throughout the whole kingdom.*

Mr. Arthur Young, in his Tour in Ireland, says, "For a country so far behind us as Ireland to have got suddenly so much the start of us in the article of roads, is a spectacle that cannot fail to strike the British traveller;" and, in speaking of the law to which this was owing, he says, "The original act was passed but seventeen years ago, and the effect of it in all parts of the kingdom is so great, that I found it perfectly practicable to travel upon wheels by a map. I will go here, I will go there; I could trace a route upon paper as wild as fancy could dictate; and every where I found beautiful roads, without break or hinderance, to enable me to realise my design.

* By 5 Geo. 3. c. 14., 1765, all former road acts from 11 James 1., in 1614, to 3 Geo. 3. c. 8., inclusive, were repealed, and the statute labour, or six days' system, abolished.

"What a figure would a person make in England who should attempt to move in that manner, where the roads, as Dr. Burn has very well observed, are almost in as bad a state as in the time of Philip and Mary.

"Arthur French of Moniva was the worthy citizen who first brought this excellent measure into Parliament. Before that time the roads, like those of England, remained impassable under the miserable police of the six days' labour.

"Similar good effects would here flow from adopting the measure which would ease the kingdom of a great burden in its public effect absolutely contemptible." *

One of the greatest efforts which has been made in modern times by the legislature, to afford, on an extensive scale, to a part of the public, the benefit of improved communication, is the plan that was adopted in 1803 for making roads in the Highlands of Scotland. Commissioners were appointed in that year for making these roads. The expense was defrayed in equal portions by grants of parliament and local contributions. The operations were conducted by Mr. Telford; and the result has been the construction of 920 miles of road in every respect suitable to the country, and of 1,117 bridges. These roads traverse the Highlands of Scotland in all directions; and, although the whole region consists of high mountains, the

* A. Young, Tour in Ireland, vol. ii. p. 56.

lines have been laid out with so much science, that the inclinations are every where moderate.

Next to tracing these roads, the principal merit consists in forming and draining them in such a manner as to place them out of the reach of all injury from the torrents to which they would otherwise be exposed.

In the districts between Glasgow, Cumbernauld, and Carlisle, upwards of 150 miles of new lowland roads have been made by Mr. Telford whilst acting under the same commission. But it was not until Mr. Telford was employed by the Commissioners appointed by Parliament, in 1815, for improving the Holyhead road, that he had an opportunity of carrying into execution a plan of road-making suitable to a great traffic on perfect principles. In that year, a sum of money having been voted by Parliament for the improvement of the Holyhead road, Mr. Telford was consulted by the commissioners with respect to the best plan of accomplishing the object Parliament had in view. He strongly recommended that, if employed by them, he should be allowed to execute all the new works upon this line of road in a substantial and perfect manner, in consequence of its great importance as the main communication between England and Ireland.

The commissioners having adopted Mr. Telford's advice, and Parliament having continued to grant further sums of money, eighty-two miles of new road have been made by him through North Wales, between Chirk and Holyhead: three miles between Chirk and the village of Gobowen, near Oswestry, and seven miles on the Holyhead and Chester road. Thirty-one miles have also been made by Mr. Telford at various places on the Holyhead road between London and North Wales, with money advanced to the Parliamentary Commissioners, on loan, by the Commissioners for giving Employment to the Poor.* These roads have been constructed in the most substantial manner. A foundation of rough pavement has been made as a bed to support the surface materials. They are uniform in breadth and superficial convexity. They are completely drained, and when carried along the face of precipices, protected by strong walls. They are acknowledged, by all persons competent to form a correct judgment on works of this kind, to be a model of the most perfect road-making that has ever been attempted in any country.†

* For a description of the works of the Parliamentary Commissioners on the Holyhead Road see the Report of a Select Committee of the House of Commons in 1830, in Appendix, No. 4.

† Extract from Mr. Telford's first Annual Report on the Holyhead Road, dated May 4, 1824, p. 17.:—

"SHREWSBURY TO HOLYHEAD.

"This portion of the great Irish road having been originally constructed in a very imperfect manner, was, till within the last five years, one of the worst roads in the kingdom. Through North Wales, in particular, no attention whatever had been paid to the essential points of a good road; it was narrow and crooked, hills had been passed over, and valleys were crossed without

The obvious utility of a work explaining the principles on which roads should be made, and

---

any regard to inclinations: no solid foundation was prepared; a very superficial coating of very bad stones or gravel was all that covered the soil; the transverse sections were often just the reverse of what they ought to be; the draining was miserably defective, and either no protecting fences, or very weak ones, existing along steep hill-sides and tremendous precipices.

"On this district there were no less than seven distinct Trusts; the revenue arising from the tolls being very limited, the trustees could not afford to employ persons whose education and previous pursuits qualified them to act as surveyors. The consequence was, that the road got unto unskilful hands, and its state of repair was just as bad as the principle of its construction.

"The increasing importance of this line of communication at length attracted the attention of Parliament. I was directed to make a survey of it in 1810; and, it having been satisfactorily shown to the successive Committees of the House of Commons, that the country through which the road passed did not in itself possess the means of providing funds for effecting any essential improvement, an Act of Parliament was passed in 1815, empowering commissioners, therein named, to expend the sum of 20,000*l.* in making such alterations as they might deem expedient.

"Under the power of this Act (the 55th Geo. 3. c. 152.) the commissioners commenced their operations in the autumn of 1815; and according as further grants were from time to time voted by Parliament, the road progressively assumed its present character. Those parts which had been the most inconvenient and dangerous have been changed to perfect specimens of what roads ought to be; steep declivities have been reduced to perfectly easy inclinations; and narrow, crooked, ill-protected portions have been converted into broad, safe, smooth, and well-constructed roads.

"The value of these improvements was felt and appreciated; and it became of the highest importance to preserve them in a perfect state, by providing an efficient system of management.

containing an illustration of those principles by a reference to the plans, specifications, and contracts

---

"By the Act of the 55th Geo. 3. c. 152. the new pieces of road, when completed, were to be made over to the local trustees, to be by them repaired and maintained. But the local Acts were imperfect; the old tolls too low; every Trust deeply in debt, and the mode of management not so perfect as it ought to be. Under these circumstances, it was thought advisable to apply to Parliament for an Act to secure to the public the lasting benefits of those improvements, by placing them under the care of one Board of Commissioners.

"Accordingly, in May 1819, the Act of the 59th Geo. 3. c. 30. was passed, in which six Trusts between Shrewsbury and Bangor were consolidated into one, and vested in fifteen commissioners therein named. The operations of this Act commenced from the first day of August following; and from that period a totally new system has been adopted on the whole line of road. At the first meeting of the commissioners they appointed a professional engineer as their general surveyor, also a clerk and a treasurer, and fixed upon a plan of management, of which the following is an outline.

"The total distance from Shrewsbury to Bangor Ferry, being 85 miles, was divided into three districts; the first, being 23 miles, extending from Shrewsbury to the boundary between Shropshire and Wales, at Chirk Bridge; the second, of 30 miles, from Chirk Bridge to Cernioge; and the third, of 32 miles, from Cernioge to Bangor Ferry.

"Over each of these districts an assistant surveyor or inspector was appointed, care being taken to select these officers from good practical workmen. Under these inspectors, a working foreman was placed on every four or five miles, with such a number of labourers under his charge as were sufficient for maintaining the road in proper repair.

"It was ordered that the labourers should be, as much as possible, employed by task, in quarrying rock, gathering field stones, getting gravel, breaking stones, scraping the road, which have been made use of in constructing the new road in North Wales, through a country persenting every kind of difficulty, has suggested the present publication. The object of it is to point out, in a clear and concise manner, the best method of tracing and constructing roads, under every variety of circumstances; and it is confidently expected, that the course which has been pursued of proceeding on experience by inserting in this work the identical plans, specifications, and contracts by which so great an extent of perfect road has been successfully made, will be found to have attained this object.

---

loading materials into carts, and all works that are reducible to measure.

"The duties of the general surveyor and clerk were, to go along the line every four weeks; the surveyor to examine the practical operations, settle all accounts with each inspector, and give the clerk a certificate, showing all the money due. The clerk to collect the tolls, and to pay every one what appeared to be owing by the surveyor's certificate, and lodge the balance of his receipts with the treasurers, Messrs. Beck and Co. of Shrewsbury."

CHAPTER I.

RULES FOR TRACING THE LINE OF A NEW ROAD.

This business of tracing the line of a road should never be undertaken without the assistance of instruments; and all local suggestions should be received with extreme caution.

To guard against errors in this important point, it is essentially necessary not to trust to the eye alone, but in every case to have a survey made of the country lying between the extreme points of the intended new road. For this purpose an experienced surveyor should be employed to survey and take the levels of all the various lines that, on a previous perambulation of the country, appear favourable. It is only by such means that the best line can be determined. These surveys should be neatly and accurately protracted and laid down on good paper, on a scale of sixty-six yards to an inch for the ground plan, and of thirty feet to an inch for the vertical section.

The map should be correctly shaded, so as to exhibit a true representation of the country, with all its undulations of high grounds and valleys, streams and brooks, houses, orchards, churches, and ponds of water adjacent to the line of road; and all other conspicuous objects should also be laid down in the map. A vertical section should be made, and the nature of the soil or different strata, to be ascertained by boring, over which each apparently favourable line passes, should be shown; for it is by this means alone that it can be determined and calculated at what inclinations the slopes in cuttings and of embankments will stand. If it be necessary to cross rivers, the height of the greatest floods should be marked on the sections; the velocity of the water, and the sectional area of the river, should be also stated.

If bogs or morasses are to be passed over, the depth of the peat should be ascertained by boring; and the general inclination of the country for drainage should be marked.

All the gravel-pits or stone quarries contiguous to the line should be described on the map, with the various roads communicating with them; and the existing bridges over the streams or rivers which are immediately below the proposed point of crossing them should be carefully measured, and the span, or waterway, stated on the section.

These preliminary precautions are absolutely necessary to enable an engineer to fix upon the best line of road, with respect to general direction, and longitudinal inclination. Without the unerring guide of actual measurement and calculation, all will be guess and uncertainty.

It may be laid down as a general rule, that the best line of road between any two points will be that which is the shortest, the most level, and the cheapest of execution: but this general rule admits of much qualification; it must, in many cases, be governed by the comparative cost of annual repairs, and the present and future traffic that may be expected to pass over the road. Natural obstructions also, such as hills, valleys, and rivers, will intervene, and frequently render it necessary to deviate from the direct course.

HILLS.

In every instance of laying out a road in a hilly country, the spirit-level is essentially necessary to show the proper line of road to be selected. The general rule to be followed in surveys is to preserve the straight line, except when it becomes necessary to leave it to gain the rate of inclination that may be considered proper to be obtained, without expensive excavations and embankments. When a deviation is made for this purpose, it becomes necessary to proceed in a direct line from a new point.

Thus, for instance, if it be decided to have no greater rate of inclination than 1 in 35, on a new line of road, from A to B (Plate I. Fig. 1.), and the surveyor, when he arrives at the point *a*, finds a greater inclination than this, he must incline from the direct line to *b*. Having then gained the summit of the hill, he does not endeavour to get back into the original straight line A B, but pursues the direct line *b* B, unless he is again obliged, from a similar cause, to deviate from it. This part of the survey being accomplished, it will then become necessary to examine the practicability of making a direct line of road between A and *b*, instead of going to the point *a*.

When hills are high and numerous, it sometimes appears, from a perambulation and inspection of the country, to be advisable to leave the straight line altogether from the beginning, in order to cross the ridges, at lower levels, by a circuitous course, in the way represented by the dotted lines A *c d*, in the above figure.

It constantly happens that although inclinations which do not exceed the prescribed rate can be had without quitting the straight line, the ridges may be crossed, at many feet of less perpendicular height, by winding the road over lower points of them; but the propriety of doing so will depend upon the length by which a road will be increased by going round to avoid passing the ridges in the direct line. The saving of perpendicular height to be passed over by a road, though a matter of so much importance and practical utility, has not hitherto received that attention from engineers which it deserves. For this reason it has been deemed advisable to bestow much consideration upon it; but, as the investigation requires minute and extensive details, it has been transferred and given in note A.

When expeditious travelling is the object, the rate of inclination that never should be exceeded in passing over hills, if it be practicable to avoid so doing, is that which will afford every advantage in descending hills, as well as in ascending them.

For, as carriages are necessarily retarded in ascending hills, however moderate their inclinations may be, if horses cannot be driven at a fast pace in going down them, a great loss of time is the result. This circumstance is particularly deserving of attention, because the present average fast rate of coach driving over any length of road can be accomplished in no other way than by going very fast down the hills. But when the hills are very steep, and a coachman cannot keep his time except by driving very fast down them, he exposes the lives of his passengers to the greatest danger.

Few travellers by stage coaches are aware of the risk they run of losing their lives in descending hills. A coachman must be thoroughly well skilled in his business, naturally cautious, and at all times sober. The wheel horses must be not only well trained to holding back, but very strong. If a pole breaks, or a pole-piece or a haime, haime strap, or drag chain, when a heavily-laden coach is descending a steep hill, at a rate exceeding six miles an hour, an overturn is almost inevitable, by reason of the coach overpowering the horses. Hence it is that ninety-nine out of every hundred coach accidents which happen, are on hills. Nothing is more important for all turnpike trustees to pay immediate attention to, as the reducing of all hills to inclinations of at most 1 in 24.

How much time is lost in descending steep hills will appear from the following hypothesis:—Suppose a hill to be so steep as not to admit of a stage coach going faster down it than at the rate of six miles an hour, five minutes will be required for every half mile: but, if the hill were of an inclination of 1 in 35, it may be driven down with perfect safety at the rate of twelve miles an hour; at which rate the time for going half a mile would be two minutes and a half, so that there is a loss of half a mile in distance for every half mile down the steep hill.

Besides the loss arising from the additional horse-power required to draw a carriage over very steep hills, there are other circumstances which make it desirable to avoid them.

In descending them, the drag becomes indispensably necessary. In coach travelling, the stopping to put it on and take it off will be the loss of at least one furlong to a coach travelling at the rate of ten miles an hour; for in slacking the pace of the horses, putting on the shoe, taking it off, and getting the horses again into their regular rate of going, nearly one minute will be occupied.

Even with the drag, heavily laden carts are always taken by their drivers into the side channels of the road to check their speed; and thus the channels are cut into deep ruts, or rather troughs, and the under-drains broken in, unless strong posts of wood or stone are set up, which are unsightly, and dangerous to other carriages descending at a quick rate. Various plans have of late been proposed for the better applying of drags to carriages, but the only effectual plan for preventing accidents is to lower all hills to rates of inclination not exceeding 1 in 24.*

---

* Extract from a Letter from Sir J. Robison, Secretary of the Royal Society of Edinburgh, respecting the French drag.

"9, Atholl Crescent, Edinburgh, 5th Nov. 1836.

"Since I had the honour of an interview with you, I have passed twice between Edinburgh and Newcastle in the Chevy Chace coach, to which the French drag has been applied during a considerable period.

"I paid close attention during these journeys to every circumstance having any relation to the working of this apparatus, and I have been led to infer that important advantages would arise from the general adoption of this drag for the public conveyances of this country.

"1st. A saving of all the time now lost in applying and removing the shoe.

"2d. A gain of time, with some relief to the horses, by taking advantage of a portion of the impetus acquired in going down one hill to assist in overcoming its opposite ascent, instead of being obliged, as at present, to strain the horses in checking the speed, and in backing the coach off the shoe.

"3d. In obtaining a command over the horses, by the power of increasing the resistance by a few turns of the winch."

This drag has been applied to coaches on other lines of road in Scotland with complete success.

Messrs. Makepiece and Pearson have invented another kind of drag, called the Patent Retarder: this has been for some time applied to the Devonport mail, and is considered by competent judges to answer effectually.

An inclination of 1 in 35 is found by experience to be just such an inclination as admits of horses being driven in a stage coach with perfect safety, when descending as fast as they can trot; because, in such a case, the coachman can preserve his command over them, and guide and stop them as he pleases. A practical illustration that this rate of inclination is not too great, may be seen on a part of the Holyhead Road, lately made by the Parliamentary Commissioners, on the north of the city of Coventry, where the inclinations are at this rate, and are found to present no impediment to fast driving, either in ascending or descending. For this reason it may be taken as a general rule, in laying out a line of new road, never, if possible, to have a greater inclination than that of 1 in 35. Particular circumstances may, no doubt, require a deviation from this rule; but nothing except a clear case that the circuit to be made in order to gain the prescribed rate, would be so great as to require more horse labour in drawing over it than in ascending a greater inclination, should be allowed to have any weight in favour of departing from this general rule. On any rate of inclination greater than 1 in 35, the labour of horses, in ascending hills, is very much increased. The experiments detailed in Mr. Telford's Seventh Report of the Parliamentary Commissioners of the Holyhead Road, made with a newly invented machine for measuring the force of traction required to draw carriages over different roads, fully establish this fact.*

The following is a table of the general results of the experiments made with a stage coach, on the same description of road; but on different rates of inclination, and with different rates of velocity:—

* See Appendix, No. 1., for a description of this machine.

Mr. Telford, in his Report to the Parliamentary Commissioners of the Holyhead and Liverpool Roads, speaking of this instrument, states, "I consider Mr. Macneill's invention, for practical purposes on a large scale, one of the most valuable that has been lately given to the public."

Mr. Babbage, the Lucasian Professor of Mathematics in the University of Cambridge, in his valuable and well-known work on the Economy of Machinery and Manufactures, in considering the injury which roads sustain from various causes, states, "As connected with this subject, and as affording most valuable information upon the points in which, previous to experiment, widely different opinions have been entertained, the following extract is inserted from Mr. Telford's Report on the state of the Holyhead and Liverpool Roads. The instrument employed for the comparison was invented by Mr. Macneill, and the road between London and Shrewsbury was selected for the place of experiment. The general results, when a waggon weighing 21 cwt. was used on different sorts of roads, are as follow:—

1. On a well-made pavement the draught is - - 33 lbs.
2. On a broken stone surface, or old flint road - 65
3. On a gravel road - - - - - - 147
4. On a broken stone road, upon a rough pavement foundation - - - - - - - 46
5. On a broken stone surface, upon a bottoming of concrete formed of Parker's cement and gravel 46"

Mr. Macneill has improved this instrument by making the machinery of it register the force of traction, the inclination of the road, and the space travelled over.—See Appendix, No. 1., p. 344, for a description of the new instrument.

Rates of Inclination.	Rates of Travelling.	Force required.
1 in 20	6 miles per hour	268 lbs.
1 in 26	6 - -	213 lbs.
1 in 30	6 - -	165 lbs.
1 in 40	6 - -	160 lbs.
1 in 600	6 - -	111 lbs.
1 in 20	8 miles per hour	296 lbs.
1 in 26	8 - -	219 lbs.
1 in 30	8 - -	196 lbs.
1 in 40	8 - -	166 lbs.
1 in 600	8 - -	120 lbs.
1 in 20	10 miles per hour	318 lbs.
1 in 26	10 - -	225 lbs.
1 in 30	10 - -	200 lbs.
1 in 40	10 - -	172 lbs.
1 in 600	10 - -	128 lbs.

From this table it will be seen that the power required, when the velocity is ten miles an hour, to draw a carriage up an inclined plane which rises one in twenty, may be taken at nearly three times as much as is required to draw it on a level road.

The note D. in the Appendix contains a mathematical explanation of the effect of hills in drawing carriages over them.

Dr. Lardner, in the evidence which he gave before the Committee of the House of Commons that was appointed in the session of 1836 " to investigate the subject of tolls and turnpike trusts in Great Britain and Ireland," made some observations on the effects of hills, which are highly deserving of attention. The following is an extract from his examination.

" A road ought to be as short as possible, consistently with some regular principle as to hills, ought it not?—Yes. Now, with respect to the acclivities, there is a distinct mechanical character which attaches to acclivities, depending on their steepness. One acclivity is not more injurious than another in the mere ratio in which it is more steep than another. There are some acclivities which afford a certain compensating effect in the descent; there are others that never fully compensate for the power lost in their ascent. There is an acclivity, or an inclination, which we designate in the department of mechanical science that relates to these things by the term of the 'angle of repose;' it is the steepest acclivity down which the carriage will not roll of its own accord—down which it will not roll by its own gravity. On more steep acclivities the carriage will roll down without any tractive power; every acclivity under that limit which will require more or less of tractive force downward. Now, acclivities which are less steep than the angle of repose give a compensation in descending for the excessive tractive force they require in ascending; that is the case with acclivities between the perfect level and the angle of repose; and I take it that that inclination should be the major limit which ought to be imposed to hills, as they are called, upon the first class of turnpike roads; the more they are under that inclination of course the better, but certainly they should never exceed it.

" Can you state that acclivity in figures?—That

will depend upon several circumstances; it will depend in some measure on the carriage, because a carriage of one structure will roll down a hill, when a carriage of another structure would not. Then it will depend upon the surface of the road; but if we take the very best class of broken-stone road surface, constructed in the best manner, so as to be as hard as can be, and a good class of carriage rolling upon it, I suppose, at a rough estimate, one in forty would be the angle of repose. I should advise the great roads not to be more steep than one in forty."

Hilly ground is not always to be avoided, as being unfit for a road; for, if the hills are steep and short, it will often be easy to obtain good inclinations, or even a level road, by cutting down the summits, and laying the materials taken from them in the hollow parts. But this must be regulated by the expense to be incurred, which is a main consideration, always to be scrupulously attended to before an engineer decides upon the relative merit of several apparently favourable lines. A perfectly flat road is to be avoided, if it is not to be raised by embanking at least three or four feet above the general level of the land on each side of it, so as to expose the surface of it fully to the sun and wind; for if there is not a longitudinal inclination of at least 1 in 100 on a road, water will not run off; in consequence of which, the surface, by being for a longer time wet and damp than it otherwise would be, will wear rapidly away, and the expense of maintaining it in order by scraping and laying on materials will be very much increased.

The great fault of most roads in hilly countries is that after ascending a considerable height, they constantly descend again before they gain the summit of the country which they have to traverse. In this way the number of feet actually ascended is made many times more than would be the case were no height, once gained, lost again.

An instance of this defect is observable in the line of old road across the island of Anglesea, on which a horse was obliged to ascend and descend 1283 perpendicular feet more than was found necessary by Mr. Telford, when he laid out the present new line, as shown by the annexed table: —

	Height of summit above high water.	Total rise and fall.	Length.	
			Miles.	Yards.
Old road -	339	3,540	24	428
New road -	193	2,257	21	1,596
Difference -	146	1,283	2	592

Another instance may be observed in the road from South Mimms to Barnet. The old road ascends three rather steep and long hills, while the new road avoids, almost entirely, two of these hills, at the same time that it is shorter by 638 yards.

In tracing a road across a deep valley between two hills, it should be carried in a direction opposite to the fall of the valley, as by so carrying it, that is, by crossing the valley at the highest practicable point, the descent and ascent are diminished.

Thus, in going from A to B, across a valley, if it be found by levelling that in a straight line the valley is too deep to make an embankment at a reasonable expense, the surveyor should try a line, A C B, higher up the valley, rather than in the direction A D B, where he would get into a lower level (see Plate I. Fig. 2.). Although this ought to be the general principle, instances may occur, where a valley may be crossed with more advantage down stream; as, for instance, if the sides of a valley contract considerably, it may require much less embankment to raise the road to the same height, than if it were carried higher up the valley; see Plate I. Fig. 3., by which it appears that it would be more advisable to take the line A D B, than either the straight line A B, or the line A C B, higher up the valley.

Another instance where a valley may be crossed with more advantage down stream, is where detached or insulated hills are situated in the valley below the straight line of direction, as represented in Plate I. Fig. 4. Here it would be proper to pass the valley lower down, to take advantage of the intervening high ground, as will be seen by the section, in which it is evident that much less embankment will be required in the line A D B, than in either the direct line A *b* B, or the line A *c* B, higher up the valley. Lately, when it was proposed by the Parliamentary Commissioners of the Holyhead Road, to improve the valley of the Geese Bridge, between Towcester and Daventry, on the road from London to Birmingham, six different surveys and plans of doing so were made. The report on these surveys is given in note A, more fully to explain the rules for crossing valleys. In many situations, particularly in mountainous countries, it will be found necessary to pass valleys or deep ravines by means of high arches of masonry, as in some parts of Scotland, where Mr. Telford has erected several great works of this description; of these, the most remarkable are the bridges over the Mouse Water, at Cartland Craigs, on the Lanark Road, represented in Plate I. Fig. 5. The bridge over Birkwood Burn, near Lesmahago, on the Glasgow Road, represented in Plate I. Fig. 6., and the Fidlor Burn Bridge, on the Lanark Road, represented in Plate I. Fig. 7.*

The suspension bridge over the Menai Straits, in North Wales, is of a similar character, for, besides its use in passing these straits, it has improved the road by its being no longer necessary

* To this list may be added the Dean Bridge over the Water of Leith at Edinburgh, which is above 100 feet high, and consists of four arches of 90 feet span; and a bridge at Pathhead, on the Coldstream road, of five arches of 80 feet span.

to descend to the level of the water. See Plate I. Fig. 8.

In most cases, however, valleys may be crossed by high embankments of earth, such as the chalk hill embankment near Dunstable, and that near Chirk, in North Wales.

In some situations it may be advisable to pass through a hill by means of a tunnel, instead of by deep cutting.

There are three works of this kind on the Simplon Road. One of them, "*la grande galerie de Gondo*," is 240 yards in length, 8½ in breadth, and the same in height. There is a similar work at Puzzuoli, near Naples, which is nearly half a league long; it is fifteen feet broad and as many high.

RIVERS.

The peculiar circumstances of a river may render it necessary to deviate from a direct line in laying out a road.

A difficulty may arise from the breadth of the river requiring a bridge of extraordinary dimensions, or from the land for a considerable distance on the sides of the river being subject to be covered with water to the depth of several feet in floods.

In these cases it may appear, upon accurately calculating and balancing the relative inconvenience and expense of endeavouring to keep a straight line and of taking a circuitous route, that upon principles of security, convenience, and expense, the circuitous course will be the best.

In general, rivers have been allowed to divert the direct line of a road too readily. There has been too much timidity about incurring the expense of new bridges, and about making embankments over flat land to raise the roads above the level of high floods.

These apprehensions would frequently be laid aside, if proper opinions were formed of the advantages that arise from making roads in the first instance, in the shortest directions, and in the most perfect manner. If a mile, half a mile, or even a quarter of a mile of road be saved, by expending even several thousand pounds, the good done extends to posterity, and the saving in annual repairs and horse labour that will be the result will, before long, pay off the original cost of the improvement.

BOGS AND MARSH GROUND.

The elastic nature of all bogs and marshes, and of all boggy and bottom land, makes it impossible to form a road of perfect hardness over a soil of this kind, unless a great deal of labour and expense is applied in draining the soil, and afterwards compressing it, by loading it with large quantities of earth embanked upon it, in order to destroy the elasticity of the subsoil.

Although the surface coating of a road over such a subsoil may be made with a great abundance of the hardest materials, and be perfectly smooth, the porous and moist texture of the subsoil will cause the road to yield to a carriage passing over it; and thus, by destroying the momentum of it, add greatly to the labour of the horses in drawing it.*

For this reason it will generally be prudent to deviate from the direct line in laying out a new road, if by doing so this sort of subsoil can be avoided, without adding much to the length of the road. But when the additional length of the road would be considerable, it will then be necessary to incur the expense of a proper drainage, and of forming a high embankment, as to compress and harden by its weight the moist and porous subsoil. Such an embankment, of 1,740 yards in length, having this object in view, was made over Maldreath Marsh, in the Island of Anglesea, on the new line of the Holyhead Road.

MATERIALS.

It will sometimes happen that road materials can be better obtained by carrying a line of road in one direction than in another. This will be a good reason for making a road deviate from the direct line, because the expense of making and repairing it will much depend on the distance which materials have to be carried.

EXPOSURE.

It is necessary, in making a road through a hilly country, to take particular care to give it a proper

* The reference which will be made to the laws of motion in a subsequent chapter will show how extremely injurious elasticity is in increasing the labour of horses.

aspect. It is a great advantage to have a road on the north side of a valley fully exposed to the sun. For the same reason, all woods, high banks, high walls, and old fences ought to be avoided, in order that the united action of the sun and wind may have full power to produce the most rapid evaporation of all moisture. Too much attention cannot be bestowed on this object, in consequence of the effect of water in contributing to cut and wear down the hardest substances. It is for this reason that road materials, when they are wet or damp, wear rapidly away under the weight and pressure of heavy carriages. The hardest limestones wear away very quickly when wet, and all stones of an aluminous character, and also gravel that consists of flint, sandstone, or other weak pebbles.

The great advantage of having a road perfectly exposed to the action of the sun and wind, will be more accurately conceived, by reference to writers of science on evaporation. Dr. Halley states, that one tenth of an inch of the surface of the sea is raised per diem in vapour. He also says, that the winds lick up the water somewhat faster than it exhales by the heat of the sun. Other writers say, the dissipation of moisture is much accelerated by the agency of sweeping winds, the effects being sometimes augmented five to ten times.

Trees are particularly injurious, by not allowing the sun and wind to have free action on the surface of roads in producing evaporation. Besides the benefit which a road receives from drying rapidly, by an open exposure to the atmosphere, there is another of great importance, namely, that of affording to horses the advantage of free respiration; for it is well known that the power of a horse to perform work with ease, particularly when moving rapidly, depends upon the quantity of cool and fresh air that he can pass through his lungs. If the cause of horses tiring or becoming ill under their work be carefully examined into, it will often be found that it is not that their muscles or limbs fail them, but their wind; and therefore it is particularly important to have a road so circumstanced, that a horse may, on all parts of it, have the benefit of a free current of air.

It may sometimes be proper to make a road deviate from a straight line, in order to go through a town; but the expediency of such a deviation must wholly depend on the general object of the road. If it be intended to expedite the communication between two places of great trade, or otherwise of great importance, then nothing can be more erroneous than allowing the general line of road to be taken from the best and shortest direction in order to pass through a town. It is for this reason that little attention should be paid to the opposition of inhabitants of towns to new roads, to be made for the advantage of the general communication between distant and important parts of the kingdom.

Some persons may be disposed to say, that a road should be made to deviate from a direct line in order to avoid crossing parks, or demesnes, and, to a certain extent, no doubt it should; but this motive ought not to be allowed to have much weight, where the consequence is to force the road over an inconvenient ascent, or to add very materially to its length. It should be recollected, that, by judicious management, a road may be made, if not ornamental, at least not injurious or detrimental to the appearance or privacy of a park, by carrying it in hollow ground, or between sunk fences.

The principle of protection of private property is itself founded on the same principle that should govern the line of a road, and that principle is the public advantage; and therefore it should be laid down and acted upon as a general maxim, that private considerations ought in all cases to be made to give way, with respect to roads, to public convenience. "For let it be remembered that society is formed for the mutual and general benefit of the whole; and it would be a very unjust measure to incommode the whole merely for the convenience, or perhaps the gratifying of the whim or caprice of an individual."*

After fixing upon the general line of a road with respect to its direction, the precise line of it must be marked out, according to the smaller acclivities and declivities of the natural surface of the country it is to pass over, in such a manner that the cuttings shall furnish sufficient earth for the embankments. As moderate curves add but little to the length of a road, they will not be objectionable, if they assist the inclinations and save expense.

* Bateman on Roads, p. 122.

CHAPTER II.

PRINCIPLES OF ROAD-MAKING.

In this chapter, the general principles will be examined, according to which the art of constructing a road should be practised; and the particular methods will be explained, by which various kinds of roads should be constructed. The art of road-making, like every other art, must essentially depend on its being exercised in conformity with certain general principles, and the justness of these principles should be rendered so clear and self-evident as not to admit of any controversy.*

It has been stated in the introduction, that the business of road-making has hitherto been almost entirely under the management of persons ignorant of the scientific principles by which it ought to be conducted. It is for this reason particularly necessary to show that it is a business of science. Until it shall be so considered, the numerous de-

* "A knowledge of true principles is indispensably necessary in every art, and in that of making roads as much as in any other. Some preliminary species of knowledge is very necessary in every superintendant or surveyor. A beaten track of knowledge is but a bad guide in cases which very frequently occur, when, amongst several ways, the best is to be preferred."—*M. S. Haldimand*, Secretary to the Bailiwick of Yverdun, on the Construction of Highways.

fects of the roads will not be observed and acknowledged, and, consequently, will not be remedied. The following extract from the examination of Dr. Lardner, before the Committee already mentioned, contains the opinion, on this point, of a man of science, in every way qualified to explain correctly the business of road-making.

" Is it a matter requiring much science and skill to arrange a road with reference to these objects?—It is quite evident it requires a very unusual combination of scientific and practical knowledge. It is obviously impracticable to make a road which would be theoretically perfect; and therefore there arises an extremely delicate inquiry as to the best possible compromise which can be made between all the inevitable imperfections, the existence of which we are forced to admit. A road, to be theoretically perfect, should be, first, perfectly straight; secondly, perfectly level; thirdly, perfectly smooth; and, fourthly, perfectly hard. If it possessed all these qualities in absolute perfection, the consequence would be it would require no tractive power at all. An impulse given to a load at one end would carry it to the other by its inertia alone. This is the ideal limit to which it is the business of a road-maker to approximate as nearly as he can, all practical circumstances being considered.

" It would appear, that the construction of a road requires a considerable degree of science and practical skill on the part of those who undertake it?—I do not know that I could suggest any one problem to be proposed to an engineer, that would require a greater exertion of scientific skill and practical knowledge, than laying out the construction of a road. Unfortunately the original laying out of a road is an employment that is rarely submitted to an engineer; he is generally controlled by circumstances. The early road-makers were almost always obliged to follow our old horse-paths in the country, in a very great degree. To lay out and design a road between two points, the surface of the country should, in the first instance, be accurately ascertained; the engineer should make himself as well acquainted with the undulations and the surface of the country as if he had passed his hand over every foot of it; and, even supposing he has a model of it before him, it becomes an extremely delicate and difficult problem to say what will be the best course to take for a line of road joining two points; he of course must encounter the undulations in such a manner as to adapt his cuttings to his embankments; that is, where he cuts through an eminence he must take care so to arrange the course of the road as that he shall have a hollow to fill up which will just employ the stuff he cuts out of the embankment; then the quality of the crust of the earth he must know, because it is not after he has begun to make his road that he is to discover the practical difficulties which stand in his way. In fact, it requires a considerable knowledge of geology; the stratification and the angles at which different soils will not only stand at the beginning, but the angles at which they will continue to stand, subject to all the actions of the weather.

" Then, in point of fact, it does not appear that the advantage of the science of engineering has been applied very extensively to the roads in this country?—Most decidedly not.

" With reference to general improvements, and also with reference to new roads in this country, it seems to be your opinion that the assistance of engineers should be more generally called in?—Clearly so; recourse ought to be had to the very first scientific and practical skill of the country; it requires the first civil engineers that can be found.

" If any plan were attempted by the legislature for the improvement of the roads, is it your opinion that it ought to be conducted under the management of the most experienced engineer that this country possesses?—I am quite decidedly of that opinion.

" The surveyors generally do not belong to the profession of engineers?—No."

If the opinion of Dr. Lardner is a correct one, that the business of road-making requires a very unusual combination of scientific and practical knowledge, it is not difficult to understand why the roads are still so imperfect; for certainly those persons who have had the management of them have not possessed this knowledge.

The following is an extract from the examination of Mr. M'Niel before the same Committee:—" In your former answer, you said that you thought it would be of advantage if the trustees of the roads were assisted by civil engineering; I wish to know whether road-making ought not to be considered as a branch of the science or art of civil engineering?— I am of opinion that the laying out a good line of road, in some parts of the country, is perhaps as difficult a subject as comes before an engineer; and it is quite impossible for country surveyors and land surveyors, who have not been accustomed to engineering pursuits, to run out a line of road with advantage to the public.

"Or short lines for improvements?—I know instances in which lines of road that are said to be improved are not so good as the old line of road.

"Now, in regard to the construction of a road, is it not necessary a person who undertakes to construct and make a new road should have that sort of education that makes him acquainted with the science of civil engineering?—It is quite necessary, and that is shown in France to a very great extent; and I believe wherever roads are made here by civil engineers, they bear a very different character to the roads in other parts of the country; and also that there is a saving in the wear of materials on a well-constructed road, and well laid out road. If a line of road has not rates of acclivity greater than 1 in 40, there will be 20 per per cent. saving over one that rises 1 in 20; this is a fact not generally known, but it is quite certain; that is to say, a road that has acclivities of 1 in 20 will cost 18 or 20 per cent. more than the one that has acclivity of 1 in 40.

"As a matter of course, a civil engineer looks to

the appearance his road has, as well as to the fitness to draw carriages; is it not always a rule with them to have it uniform as to breadth and shape?—Yes; there are certain rules which an engineer would always adopt, that is, a certain uniform width and a certain curvature, a certain height of footway, and a certain width of waste and fences, according to the description of road he was to make.

" To acquire that degree of uniformity, is it not necessary to use instruments, and to have that sort of habitual method of managing works that can only be acquired by a regular education?—I conceive so; I do not think a road can well be laid out, except by a professional person.

" Have you found it the practice to appoint engineers as surveyors of roads?—No; I do not know an instance of it, except on the road between Shrewsbury and Holyhead, and there the effect is very apparent.

" What class of persons are they commonly?—Generally farmers; in some instances they have been tradespeople.

" May not a great deal of what may be considered to be imperfect in the roads of this country be attributed to the want of having more assistance from the profession?—Yes; I think the fact I have stated, that a saving of 20 per cent. in repairing a road might be made in a very slight alteration of declivity in a road, will prove that principle, and also that there will be a saving when the road has the appearance of uniformity and neatness about it; for the men who put out the stone can see when it is out of shape and where it gets weak, and they instantly repair it; by this means there is not the same jolting, the same degree of resistance, to carriages passing over it; and nothing but a perfect and uniform line of curves and levels will enable the eye of a workman to see where the deficiency takes place.

"Generally speaking, the roads are not uniform as to breadth, convexity, or width?—No; there does not appear to be any system in this country on this very important point; generally speaking, no road that I am acquainted with has uniform width and height of footpath, and curvature of surface, even for half a mile in length.

"Would it contribute to the good order and keeping up of a road to pay attention to these points?—It would be a great saving.

"A road is easier kept clean and dry?—Yes, and more easily seen where it becomes weak.

"Are the workmen more attentive and careful when it is all put in a proper shape?—They would become of a different character; a workman, as soon as he got a uniform and neat road, would have a pride in this work, and would keep it in better order and free from ruts, weeds, &c."

Before proceeding to examine what are the right principles for constructing roads, it will be useful to show correctly what is the proper object to be secured in making a road. It is owing very much to omitting to take this circumstance into consideration, that vague and erroneous opinions so generally prevail respecting what a road ought to be in order that it should be a perfectly good one. Almost all road-makers who have acquired reputation with the country gentlemen who are the trustees of roads, have acquired it under the idea that smoothness was the sole object to be secured, without reference to the indispensable necessity of making a road extremely hard and solid as well as smooth. Dr. Lardner says,—" Until a comparatively late period a very prevalent, indeed almost universal error obtained with respect to roads. All that people considered was what they conceived to be an easy motion to the passengers; that which was easiest to the passengers was concluded to be also the easiest to the horses; or perhaps I should be more correct in saying the horses were not considered at all. People never thought of taking into consideration the mechanical force which was necessary to draw a load along a road. If there were two roads with surfaces equally smooth, (acclivities of an equal steepness, and along which the passengers felt themselves equally comfortable,) those roads were at once assumed to be, to all intents and purposes, equally good; a greater mistake could scarcely be found than that. Suppose a road surface were made of Indian rubber, the surface being as smooth as it can be imagined to be, no road could be worse for traction; the wheels would sink into the surface, and the tractive force would be continually pulling up a hill; it would have the effect of a continual ascent. The surface of the road should be as hard and as unyielding as art can make it; the wheel should not sink; no temporary depression should take place, even though that depression be restored by elasticity after the wheel is removed. By whatever means, this end must be attained, it is quite essential, although there may be a difference in the means of attaining it; but attained it most certainly ought to be."

"The main object of a road connecting two places is to enable loads to be transported from the one place to the other in the least possible time, and with least possible expenditure to tractive power. This tractive power depends upon several qualities in the road; first, upon its levelness; secondly, upon the smoothness of its surface; and thirdly, upon a quality which may be called hardness, the absence, in fact, of elasticity."

One of the most important and most obviously correct principles of road-making is that which requires a road to be made of a substance in due proportion to the weight and number of the carriages that are to travel over it.

But although this is, in appearance, a self-evident proposition, no rule in practice is so universally violated.

Let the construction of any turnpike road commonly considered as among the best be properly examined; that is, let the quantity of hard-road materials that compose the crust over the subsoil be measured, and it will almost universally be found that it consists of only from three to five, or at most six inches in thickness.* Whereas, instead of this weak and defective system, it may be laid down as a general rule, that on every main road where numerous heavy waggons and stage coaches heavily laden are constantly travelling, the proper degree of strength which such a road ought to have cannot be obtained except by forming a regular foundation with large stones, set as a rough pavement, with a coating of at least six inches of broken stone of the hardest kind laid upon it; and further, that in all cases where the subsoil is elastic, it is necessary, before the foundation is laid on, that this elastic subsoil should be rendered non-elastic by every sort of contrivance; such, for instance, amongst others, as perfect drainage, and laying a high embankment of earth upon the elastic soil, to compress it.

Although a road, if made with a thick bed of well-broken hard stones laid upon the subsoil, may, no doubt, be, to all appearance, a hard and a good

* See Mr. Telford's first Annual Report on the Holyhead Road, in 1823, where tables are given showing the result of trials made along the whole line of road from London to Shrewsbury of the depth of materials, by sinking holes into the road at short intervals. The average depth of materials was as follows on some of the trusts: —

Whetstone Trust	- -	4 inches.
St. Alban's ditto	- -	4 ditto.
Dunstable ditto	- -	4¼ ditto.
Puddle Hill ditto	- -	3¾ ditto.

Almost all other roads, which are commonly considered good ones, would, if similar trials were made, be found to be in the same defective state.

one, still the elasticity of the subsoil will have a considerable effect in adding to the tractive force necessary to draw carriages over it; for the subsoil will yield more or less (in proportion to the elasticity which belongs to all kinds of earth,) under the incumbent weight. It is therefore only by proceeding in the way recommended, that is, by proper drainage and pressure, and by making a foundation of large stones in the form of a regular pavement, that this elasticity can be effectively diminished; for to remove it altogether is perhaps impracticable.

Rightly to understand this principle, which requires that roads should be constructed with a body or depth of materials four or five times greater than is commonly given to them, it is requisite to illustrate and establish the grounds on which it rests, by reference, first, to the laws of science concerning moving bodies, and, secondly, to experiments, which accurately prove the force of traction on different kinds of roads.

As a carriage for conveying goods or passengers when put in action becomes a moving body, in the language of science, the question to be examined and decided is, how a carriage, when once propelled, can be kept moving onwards with the least possible quantity of labour to horses, or of force of traction?

Sir Isaac Newton has laid it down as a general principle of science, that a body, when once set in motion, will continue to move uniformly forward in a straight line by its momentum, until it be stopped by the action of some external force. This proposition is admitted and adopted by all natural philosophers as being perfectly true, and therefore, in order to apply it to roads, it is necessary to enquire what species of external force act in a manner to diminish and destroy the momentum of carriages passing over them. With respect to these external forces, the general doctrine is, that they consist of, 1st, collision; 2d, friction; 3d, gravity; and 4th, air.*

1st. The effect of collision is very great in diminishing the momentum of carriages; it is occasioned by, and is in proportion to, the hard protuberances and other inequalities on the surface of a road. These occasion, by the resistance which they make to the wheels, jolts and shocks, which waste the power of draught, and considerably check the forward motion of a carriage.

The mathematical illustration of the effect of collision in producing this resistance is given in note B.

2d. Friction has very great influence in checking the motion of a carriage; for, when the wheels come into contact with a soft or elastic surface, the friction which takes place operates powerfully in obstructing the tendency of the carriage to proceed; the motion forwards is immediately retarded, and would soon cease if not renewed by the efforts of the horses. The "resistance," Professor Leslie says, "which friction occasions, partakes of the

* See Wood's Mechanics, p. 20.

nature of the resistance of fluids; it consists of the consumption of the moving force, or of the horse's labour, occasioned by the soft surface of the road and *the continually depressing of the spongy and elastic sub-strata of the road.*" *

An ivory ball, set in motion with a certain velocity over a Turkey carpet, will suffer visible relaxation of its course; but, with the same impelling force, it will advance farther if rolled over a superfine cloth; still farther over smooth oaken planks; and it will scarcely seem to abate its velocity over a sheet of pure ice.

This short explanation of the nature and effects of collision and friction is sufficient to show, that smoothness and hardness are the chief qualities in a road. But perfect smoothness cannot be obtained without perfect hardness, and therefore the business of making a good road may be said to resolve itself into that of securing perfect hardness.

With the view of taking the right course for securing this object, the first thing a road trustee or engineer should do, is to form a correct notion of what hardness is; because the common habit of overlooking this circumstance has been the source of great error in forming opinions upon the qualities of different kinds of roads.

Gravel roads, for instance, to which an appearance of smoothness is given by scraping them, at a vast expense, and patching them with thin layers of very small gravel, are very commonly declared

* Elements of Natural Philosophy.

to be perfect, and unequalled by any other kind of road. But if the best gravel road be compared with one properly constructed with stone materials, the hardness of the former will be found to be greatly inferior to that of the latter, and the error of the advocates of smooth-looking gravel roads will be immediately made manifest.

By referring to works of science, it will be seen that hardness is defined to be that property of a body by which it resists the impression of other bodies which impinge upon it; and the degree of hardness is measured by the quantity of this resistance. If the resistance be so complete as to render it totally incapable of any impression, then a body is said to be perfectly hard.*

Now this hardness is the hardness which a road ought to have as far as is practicable, and it is the chief business of a scientific road maker to do every thing necessary to produce it. For this purpose, when making a new road, he should first select or establish a substratum of soil or earth that is not spongy or elastic, for the bed of the road; and then he should so dispose the materials of which its crust is to consist, as to form a body sufficiently strong to oppose the greatest possible quantity of resistance to the weight of heavy carriages passing over it.

That an elastic subsoil is unfit for a road is evident from the nature of the resistance occasioned by friction, as above described by Professor Leslie,

* Bridges' Natural Philosophy, vol. i. p. 150.

and from the terms of the definition of hardness; for however strong the crust of materials may be which is formed over such a subsoil, it will not be capable of opposing a perfect resistance to a heavy moving body. The moving body will sink more or less in proportion as the subsoil is elastic, and the hardness of the road will be imperfect in proportion as this sinking takes place; so that nothing can be more necessary, as a preliminary step in making a new road, than to take every possible precaution to avoid elastic subsoils, or to destroy the elasticity as much as possible, when no other can be found.

After the engineer has prepared a proper substratum of earth for the bed of a road, he must next construct a crust of materials in such a manner as, when consolidated, to possess such a degree of hardness as will not admit the wheels of carriages to sink or cut into it. For this purpose it will not be sufficient merely to lay upon the prepared bed of earth a coating of broken stones; for the carriages passing over them will force those next the earth into it, and, at the same time, press much of the earth upwards between the stones; this will take place to a great degree in wet weather, when the bed of earth will be converted into soft mud by water passing from the surface of the road, through the broken stones, into it. In this way a considerable quantity of earth will be mixed with the stone materials forming the crust of the road, and this mixture will make it extremely imperfect as to hardness, for it cannot, in fact, be perfectly hard unless it consists wholly of stones. It might be possible, in some measure, to cure this defect by laying on a succession of coatings of broken stones; but several of these will be necessary, and, after all, in long-continued wet weather, the mud will continue to be pressed upwards from the bottom to the surface of the stones. If even a coating of from sixteen to twenty inches of stones be laid on, it will produce only a palliative of the evil. So that this plan of making a road will be not only very imperfect, but at the same time very expensive.

Mr. Telford's plan of making a regular bottoming of rough, close-set pavement, which has completely succeeded on the Holyhead Road, the Glasgow and Carlisle Road, and several other roads in Scotland, is one that secures the greatest degree of hardness; it is also attended with much less expense than when a thick coating of broken stones is used, for six inches of broken stones is sufficient when laid on a pavement, and the pavement may be made with any kind of common stone.

If the stones in making the bottoming are laid with their broadest face downwards, and the interstices are filled with stone chips well driven in, the earthy bed of the road cannot be pressed up so as to be mixed with the coating of broken stones. This coating, therefore, when consolidated, will form a solid uniform mass of stone, and be infinitely harder than one of broken stones, when mixed with the earth of the substratum of the road. It is by proceeding in the way here recommended that the friction of wheels on the surface of a road will be reduced as much as possible.*

To comprehend thoroughly the great importance of making a regular and strong foundation, it should be borne in mind, that roads are structures that have to sustain great weights, and violent percussion; the same rules therefore ought to be followed with them as are followed with regard to other structures.

In building edifices which are to support great weights, whether a church, a house, or a bridge, the primary and indispensable consideration of the architect is to obtain a permanently firm and stable foundation. He well knows that unless this be first substantially made, no future dependence can be placed on the stability of the intended superstructure: but this most requisite precaution has but recently been attended to in the formation of roads, and only on those roads in Scotland, and between London and Holyhead, which have been under the direction of Mr. Telford.

If the foundation of a road be not sufficient and equal to the pressure it has to sustain, the whole fabric, though in other respects ever so well constructed, must fail in permanent stability, and its hardness will be imperfect on account of its elasticity.

Having now stated all that the rules of science relating to moving bodies suggest, in order to de-

* The mathematical illustration of the effect of friction on carriages is given in note C.

fend the principles of road-making, which have been laid down as those proper to be adopted, we shall proceed further to illustrate and support these principles, by reference to experiments of the force of traction on different kinds of roads. These experiments have been made with the machine invented by Mr. Macneill, which has been already mentioned, and which may be relied upon for their accuracy, in consequence of their having been carefully examined by several eminent civil engineers.

These experiments uniformly show, that the force of traction is, in every case, in an exact proportion to the strength and hardness of a road. The following are the results: on a well-made pavement, the power required to draw a waggon is 33 lbs.; on a road made with six inches of broken stone of great hardness, laid on a foundation of large stones, set in the form of a pavement, the power required is 46 lbs.; on a road made with a thick coating of broken stone, laid on earth, the power required is 65 lbs.; and on a road made with a thick coating of gravel, laid on earth, the power required is 147 lbs. Thus it appears that the results of actual experiments fully correspond with those deduced from the laws of science.*

* The following is an extract from the evidence of Mr. Macneill before the Committee on Turnpike Roads (1836): —

"Are you still satisfied of the principle of your machine being a correct one, as to ascertaining the draughts of carriages?— I am quite certain of its being perfectly correct, for we have tried it in some cases, and compared it with a weight hanging over a pully, and the results were the same. It is only in cer-

It has been considered necessary to enter into these details in showing that no road can be cor-

---

tain cases where a weight over a pulley could be applied; it could not be done practically over a road of any length.

"Has anything occurred as to the soundness of your principle being controverted by other engineers?—No.

"Is it generally adopted by them as a correct machine?—Yes, and referred to in very many cases. In France there has been a petition to the Chamber of Deputies, founded on my experiments, relative to the mode of ascertaining the draught of carriages, and the saving by using springs.

"You were the author of that algebraical calculation delivered in the Lords?—Yes.

"Does it correspond with the results made by the machine?—It was founded on experiments made by the machine; it was a formula that would give the power required to draw a carriage over a road in a section of that road, from data determined by experiments made by the machine.

"Is your machine calculated to give the draught on setting the body in motion, or when it is in motion?—Both.

"Then it would appear that your former calculations, as to different effects of different roads on the draughts of carriages, are correct?—Yes, quite correct; and they have been confirmed by very many experiments I have made since I was examined before the Committee of the House of Lords.

"Then, in fact, the general conclusion is, that a road is good for its object, namely, of diminishing the draught of a carriage, in the proportion that it is hard and smooth?—The great advantages of the roads appearing by the machine is certainly in proportion to their solidity and their strength, and their want of yielding. If it could be a perfectly solid mass of stone or metal, the least resistance would be presented; that is shown both on stone tramways and on metal tramways and metal rails. There are some metal tramways laid in Glasgow on rather a steep hill, and it is not at all unusual for a horse to take from two to three tons; that arises merely from the saving in the resistance of the surface friction being lessened.

rectly called a good one unless it is so constructed as to be very strong and very hard, because all the main roads of the kingdom are still very defective in respect to strength and hardness. This is a fact which cannot be disputed; first, because there is always mixed up with the body of materials, which forms the crust of every road, a great quantity of earth; secondly, because this crust is every where too thin; and, thirdly, because it very frequently lies upon an elastic substratum. Although there may be exceptions, this may be taken as an accurate description.

Notwithstanding that all the roads are now much better than at any former period, and may deserve to be called good, in comparison with those of ten or fifteen years ago; when it is considered how much better they would be if they were reconstructed with a proper foundation coated with

---

"That is, from the smoothness of the surface?—Yes, from the smoothness and hardness.

"So that if clean material of any road nine inches thick were properly beat down, that will not yield?—Nine inches will yield very much.

"What, on an old road of nine inches thick?—Yes, with heavy waggons. One of the great advantages arising from Mr. Telford's system of forming roads by large stone pavements, is from the fact that one point is distributed, that the pressure of the wheels is distributed over a large space. The wheels of the carriage rest on, say two inches of surface, but that is carried to a large pitching stone below, which rests on the soil, and the weight is distributed over a large surface at the bottom; that is to say, over a surface a foot or nine inches long, and six or eight inches wide; it is lessened very much indeed on the surface that bears on the earth."

broken stones of great hardness, they should still be set down as being imperfect. Let any trustee or surveyor who doubts this, reconstruct a mile of a road, now considered an excellent one, with a bottoming of pavement, coated with hard stones, and no stage coachman who shall drive over it will hesitate to bear testimony to the increased ease with which his horses do their work upon it.

The following extract from the examination of Dr. Lardner will serve to explain, and at the same time support, the statements contained in the preceding pages.

" *On the elasticity of roads.*

" Do you speak from experience on this point of elasticity?—Not from experience, as an engineer, but only from having devoted a good deal of time to the consideration of this subject, and being perfectly acquainted with the experiments that have been made, and the experience we have had upon roads; and I also give that opinion (that a road should not be elastic) upon general scientific principles.

" With reference to the general laws of motion? —Undoubtedly; it is not a point about which any two scientific men can differ; there can be no difference of opinion about it. I mean that that quality is best for the surface of a road which will not permit it to alter its figure under the pressure of the wheels.

" Then the degree of hardness will depend upon the degree in which elasticity is absent?—Yes, certainly. It will convey to the Committee more correctly my meaning, if I state that the quality of the road ought to be such, that, as the wheels roll over it, it should not suffer any change of its figure.

" This quality is strictly in conformity with the laws of science as relating to moving bodies?—Strictly.

" According as I understand you, there is no difference of opinion among scientific men as to the necessity of having a road as hard as possible?—Yes, in order to offer the least possible resistance to the tractive force.

" What, in your opinion, is the proper way of getting rid of the elasticity in roads when constructing them?—That is, I think, a question which involves a great deal of practical difficulty. It is quite clear that, *cæteris paribus*, the thicker the crust of the road is the harder it will be, because a thick crust will not yield as much as a thin one.

" In laying out a line of road you would avoid, if possible, going over marshy ground?—Clearly so; or at least, if I did go over it, I would take care to press it down, so as to destroy its marshy character, as we have done in the Chatmoss, on the Liverpool and Manchester railroad."

The explanation of the laws of motion, which has been given in this chapter, as applicable to the subject of road-making, and of the effect of an elastic substratum of a road in consuming the moving force, and adding to the horse's labour, is quite conclusive in showing how much at variance with the first principles of science the following doctrines are, which are to be found in some modern publications.

"That a foundation or bottoming of large stones is unnecessary and injurious on any kind of subsoil."

"That the maximum strength or depth of metal requisite for any road is only ten inches."

"That the duration only, and not the condition of a road, depends upon the quality and nature of the material used."

"That free stone will make as good a road as any other kind of stone."

"That it is no matter whether the substratum be soft or hard." *

* The passages marked with inverted commas have been extracted from the publications of Mr. M'Adam.

As many persons advocate Mr. M'Adam's doctrine of elastic roads, it may serve to show the real value of it, by putting it in juxta-position with that of the celebrated natural philosopher, the late Professor Leslie.

Extract from the Evidence of Mr. M'Adam. (*Remarks on Road-making*, p. 111.)

"What depth of solid materials would you think it right to put upon a road in order to repair it properly?—I should think that ten inches of well-consolidated materials is equal to carry any thing.

"That is, provided the substratum is sound?—No: I should not care whether the substratum was soft or hard: I should rather prefer a soft one to a hard one.

Extract from Professor Leslie's "Elements of Natural Philosophy."

"The resistance which friction occasions (to carriages) partakes of the nature of the resistance of fluids: it consists of the consumption of the moving force or of

Mr. Wingrove, an eminent practical road surveyor, observes, in a Treatise on the Bath roads, after quoting the preceding passages from Mr. M'Adam's book, "that with respect to these opinions on road-making, nothing but the complete ignorance of the public upon all matters concerning road-making could ever have suffered rules, so contrary to every thing like sound principles, to have had a single moment of favourable consideration."*

The resistance produced by gravity, in checking the progress of a moving body on a road, is little or nothing when a road is horizontal, because as gravity acts in a direction perpendicular to the plane of the horizon, it neither accelerates nor retards the motion.† But when the road is not

---

"You don't mean to say you would prefer a bog?—If it was not such a bog as would not allow a man to walk over, I should prefer it.

"But must not the draught of a carriage be much greater on a road which has a very soft foundation than on one which is of a rocky foundation?—I think the difference would be very little indeed, because the yield of a good road on a soft foundation is not perceptible." the horse's labour, occasioned by the soft surface of the road, and *the continually depressing of the spongy and elastic substrata of the road.*"

* Mr. Wingrove was for several years the surveyor of nearly all the roads in the neighbourhood of Bath. In 1825 the author accompanied him in making an inspection of them, and found the rules which Mr. Telford recommends had been most effectually acted upon throughout the whole of these roads, and that they had been brought to as high a state of improvement as the money which was allowed for them would admit of.

† Wood's Mechanics, p. 20.

horizontal, the power of gravity is a great impediment.

A mathematical illustration of the effect of gravity on hills is given in note D.

The resistance arising from the action of the air is very variable; in some cases, it acts powerfully; but as its influence is the same whether the road be bad or good, little need be here said on the subject: it will be sufficient to state, that by experiments detailed in Smeaton's Reports, it was found that the force of the wind on a surface 1 foot square was 1 lb., when the velocity of the wind was 15 miles an hour, or what would be termed a brisk gale; 3 lbs. when the velocity was 25 miles an hour, or what would be termed a very brisk gale; 6 lbs. when the velocity was 35 miles per hour, or what might be termed a high wind; and 12 lbs. to the square foot when the velocity was 50 miles an hour, or what might be termed a storm. Supposing, therefore, that the surface of that part of a carriage acted upon by the direct influence of the wind to be 50 superficial feet, the resistance it will meet from a brisk gale of wind acting against it will be about 50 lbs. when the carriage is slowly moved; but if the carriage be supposed to move directly against the wind with a velocity of 10 miles an hour, and the wind to move with a velocity of 15 miles an hour, the resistance against the carriage will amount to 3 lbs. on the square foot, or 150 lbs. on the carriage, which is fully equal to the power which two horses should be required to exert, when moving with a velocity of 10 miles an hour. From this the difficulty is evident of driving stage coaches at a rapid rate against high winds.*

* The late M. Navier, one of the most scientific members of the French Institute, and Engineer in Chief of the Administration of the Ponts and Chaussées, has borne testimony to the correctness of the principles and reasonings contained in this chapter, by having given a translation of the whole of it in his book on Roads. He says, "the large extract which I here insert contains that part of the Treatise of Sir Henry Parnell which has appeared to me to be the most interesting to French engineers, and which relates most immediately to the object in view." — *Considérations sur les Principes de la Police du Roulage, et sur les Travaux d'entretien des Routes. Par M. Navier. Paris*, 1835.

CHAPTER III.

FORMING A ROAD.

In marking out the line of a road, a great deal of expense in cutting and embanking for forming the bed on which the road materials are to be placed, may be avoided by a judicious selection of the high and low ground which the surface of the country affords.

The chief care, where a road must be carried over a high elevation, is to lay it out so that it shall not have any fall in it from the point from which it departs till it reaches the summit. The lowering of heights, and the filling of hollows, should be so adjusted as to secure gradual and continued ascending inclinations to the highest point to be passed over.

It is a most important part of the business of a skilful engineer to lay out the longitudinal inclinations of a road with the least quantity of cutting and embanking.

He must do this by measuring and calculating the quantity of earth to be removed in cuttings, and taking care that it shall exactly make the embankments for raising the hollows to the required heights; a proper allowance being made for the subsidence of the soil according to its quality, without leaving an overplus to be carried to spoil.*

When it is necessary to make a deep cutting through a hill, the slopes of the banks should never be less, except in passing through stone, than two feet horizontal to one foot perpendicular; for though several kinds of earth will stand at steeper inclinations, a slope of two to one is necessary for admitting the sun and wind to reach the road. The whole of the green sod and fertile soil on the surface of the land cut through should be carefully collected and reserved, in order to be laid on the slopes immediately after they are formed.

If a sufficient quantity of sods cannot be procured in the space required for the road, the slopes should be covered with three or four inches of the surface mould, and hay seeds should be sown on it; by this plan the slopes will soon be covered with grass, which will be a great means of preventing them from slipping.

When stones can be got the slopes should be supported by a wall raised two or three feet high at the bottom of them. These walls prevent the earth from falling from the slopes into the side channels, and add very much to the finished and workmanlike appearance of a road.

In many cases it may be advisable, particularly if an additional quantity of earth be wanted for an embankment, to make the slopes through the cut-

* See Mr. Macneill's Work on Cutting and Embanking. Published by Roake and Varty.

tings on the south side of an inclination of three horizontal to one perpendicular, in order to secure the great advantage of allowing the sun and wind to reach more freely the surface of the road.

In districts of country where stones abound, expense in moving earth and purchasing land may be avoided, by building retaining walls, and filling between them with earth. In rocky and rugged countries this is generally the best way of obtaining the prescribed inclinations.

In forming a road along the face of a precipice, a wall must be built to support it. The difficulty in such a place is not so great as is imagined, for the face of a precipice is seldom perpendicular, and an inclination of half a foot perpendicular to one foot horizontal will admit of a retaining wall being built. By building such a wall, say thirty feet high, and cutting ten feet at that height into the rock, and filling up the space within the wall, a road of sufficient breadth will be obtained, as shown in Plate II. fig. 1.

In forming a road along the face of a hill that is indented with ravines, in place of carrying it over the natural surface of the land, in order to keep it level, the projecting points should be cut through, and the level obtained by laying the earth across the hollows, as shown in Plate II. fig. 2., where the road, instead of following the sinuosities of the hill, as represented by the dotted line *a a a*, takes the line *b b b*.

In forming the bed for the materials care should be taken (except where cutting into the surface is wholly unavoidable in order to obtain the proper longitudinal inclinations,) to elevate it with earth two feet at least above the natural surface of the adjoining ground: by following this course the road will not be affected by water running under or soaking into it from the adjoining land. In arranging the inclinations, they should be obtained by embanking, when that is practicable, in preference to cutting, in consequence of the better exposure of the surface on embankments, and of the risk in all cuttings of the slopes falling down.

Almost all old roads are sunk below the adjacent fields: this has arisen from their continued wear, and from carrying away the mud. No improvement is more generally wanted than new forming these roads, so as to raise their surfaces above the level of the adjoining land. This would greatly contribute to their hardness, to economy in keeping them in repair, and to enable horses to work with the advantage of having sufficient air for respiration.

EMBANKMENTS.

Great care is necessary in making high embankments. No person should be intrusted with these works who has not had considerable experience as a canal or road maker; for, if the base of an embankment is not formed at first to its full breadth, and if the earth is not laid on in regular layers or courses not exceeding four feet in thickness, it is almost certain to slip. In forming high embankments the earth should be laid on in concave courses, as represented in Plate II. fig. 3., in order to give firmness and stability to the work. It is not at all uncommon in many parts of the country to see embankments formed convexly, as represented in Plate II. fig. 4., the consequence of which is, that they are for ever slipping.

There have been but few attempts to make embankments by turnpike trustees that do not afford illustrations of this defect, and of a want of knowledge of the proper rules by which these works should be managed. No doubt, a chief reason for making cuttings and embankments, as is frequently the case, with slopes of one to one, has been to save expense in the purchase of land, and moving earth. But the consequence of making such slopes is that the earth is constantly slipping; so that, in the end, the expense is always greater in correcting the original error, than it would have been had proper slopes been made in the first instance.

In forming embankments along the sides of hills, or what is called side-forming, the rule that should be followed is, that the slope to be covered should be cut into level steps to receive the earth, otherwise it will be very liable to slip down the hill: in such cases, the earth should be well compressed, and great care should be taken to intercept all the land springs about it by proper drainage. For this purpose, a drain should be cut on the upper side of the road, and open drains should be made on the side of the hill above the road, to catch the surface water.

The figure 5. in Plate II. explains the manner in

which the ground should be formed for side embankments, by cutting the level steps *a a a*, and shows where the drains should be made.

The slopes at which cuttings and embankments can be safely made depend entirely upon the nature of the soil. In the London and plastic clay formations, it will not be safe to make the slopes of embankments or cuttings, that exceed four feet in height, with a steeper slope than three feet horizontal for one foot perpendicular. In cuttings in chalk or chalk marl, the slopes will stand at one to one. In sandstone, if it be solid, hard, and uniform, the slopes will stand at a quarter to one, or nearly perpendicular.

If a sandstone stratum alternates with one of clay or marl, as represented in Plate II. fig. 6., it is difficult to say at what rate of inclination the slopes will stand; this will, in fact, depend upon the inclination of the strata. If the line of the road is parallel to the line of the bearing of the strata*, in such cases, large masses of the stone become detached, and slip down over the smooth and glassy surface of the subjacent bed. There are many instances of slips in sandstone and marl strata, under such circumstances as those now described, where the slopes are as much as four to one. If the road is across such strata, or at right angles to

* The line of intersection of any inclined stratum with the horizontal plane is called the line of bearing of that stratum, or the drift line. The dip, or inclination, of the stratum is the angle formed between a horizontal plane and a line drawn at right angles to the drift-line on the bed of the stratum.

the line of bearing, then the slopes may be made one and a half to one, as represented in Plate II. fig. 7; but if the strata are horizontal, even should there be thin layers of marl between the beds of stone, as in Plate II. fig. 8., the slopes will stand at a quarter to one. But it will be necessary, if the beds of marl exceed twelve inches in thickness, to face them with stone.

In the Oxford clay, which covers so great a portion of the midland counties of England, the slopes should not be less in any instance than two to one, and even in some parts of this formation they should be made three to one, if the cuttings are deep. In all such cases, if any beds of gravel or sand are found intermixed with the clay, as shown in Plate II. fig. 9., drains should be cut along the top, and even in the sides of the cuttings; for, if this precaution be not taken, the water, which will find its way into the gravel, will, by its hydrostatic pressure, force the body of clay down before it, and slips will take place even when the inclinations are as much as four to one; and when this occurs it is extremely difficult to re-establish them.

In limestone strata, if solid, slopes will stand at a quarter to one; but in most cases limestone is found mixed with clay beds, and in such cases the slopes should be one and a half or two to one. In the primitive strata, such as granite, slate, or gneiss, slopes will stand at a quarter to one.

Before quitting this subject, it is proper to remark, that in every instance of deep cutting, the greatest pains should be bestowed in examining the character of the material to be removed; much difficulty will be avoided by proceeding in this way: but, on the whole, the best general rule to follow is, to lay out a line of road so as to avoid as much as possible deep cuttings and high embankments; they are always attended with great expense, and are unavoidably liable to many objections.

The footpaths should be formed at the same time as the bed of a road; also the fences, if they consist of mounds of earth or ditches: but these will be more particularly described in a subsequent chapter.

The following directions are taken from specifications according to which parts of the Holyhead road have been made.

First Specification.

"The black line on the section represents the natural surface of the ground, in the longitudinal direction of the new line, at about the middle of the space to be occupied by the road. The red line represents the proposed finished longitudinal surface of the bed, or what the road materials are to be laid upon; the red figures denote the depths of cuttings and the heights of the embankments, and also the rates of the inclinations: these rates of acclivity are to be strictly adhered to, and it is expressly stipulated that the contractor is to satisfy himself by his own measurement, or in any way he may think proper, as to the heights and depths, or any irregularities, of other parts of the surface of the ground to be cut down or embanked, or where there is to be side-cutting and forming, as no future claim on any pretence whatever will be allowed.

"The breadth of the finished road is to be thirty-six feet; viz. thirty feet for the carriage way, and six feet for the footpath. The slopes of all embankments from the outside of the quick borders are to be two horizontal to one perpendicular, neatly dressed and covered with green sod at least four inches thick, evenly laid, and closely jointed.

"The slopes of the cuttings on the southern sides are to be three horizontal to one perpendicular, and those of the northern side to be two horizontal to one perpendicular; these slopes are to be correctly formed, neatly dressed, and covered with a good vegetable sod, the green side placed uppermost, and neatly jointed, and evenly laid, and to be at least four inches thick.

"The surface of the bed for the materials of the carriage way is to be formed level from side to side, the breadth between the bottom of the side slopes in the cuttings at the level of the bottom of the road materials is to be thirty-one feet.

"The surface of the bed for the hard materials of the footpath is also to be level, and to be seven inches above that of the carriage road. The necessary breadth will be gained by the road materials resting on the sides of the slopes.

" Where there is to be cutting in the side of a hill, the slope of the bank is to be two horizontal to one perpendicular: the embankment is to be secured by cutting the slope of the hill below the line of road into level steps to receive the earth, and the road materials are not to be laid on the embankment until the inspector is satisfied it will stand."

Second Specification.

" In each of these lots the contractors are to make the line of the road agreeably to the plans and longitudinal sections made and signed by Mr. Telford, as laid out upon the ground by him, or such person as he shall appoint. The breadth, shape, and construction shall be according to the particulars, and the cross sections for construction made out by Mr. Telford; that is to say, on level ground the bed shall be formed by removing the vegetable and other soft matters, and brought to a perfect level and consolidated state. If the ground is soft bog or morass, and less than four feet in depth, with hard ground below, the soft matter shall be removed; but if a greater depth, the whole surface shall be covered with two rows of swarded turf, the one laid with its swarded face down, and the other upwards. Where the road is formed on sloping, it shall either be cut for the whole breadth into the solid bank, with as much more as to afford a solid foundation for a fence wall, or as much shall be cut from the upper side as shall bring the lower to a proper level. If this consists of loose soil, it must be compressed by means of water, or shall be left through a part of the winter to receive the snows and rains; but no soft, boggy, or peat substance is on any account to be laid behind the retaining walls. Where the cutting on the upper side consists of rock rubbish, gravel, or mountain clay, it will only require to be properly levelled as the work is carried on."

The following specification has been successfully acted upon in forming a road over a peat bog in Ireland: —

"When the line of the road has been traced out to the exact width and line of direction, main drains are to be cut on each side eight feet wide at top, four feet deep, and eighteen inches wide at bottom; the peat dug out of these drains is to be spread over the surface of the roadway in form of a ridge, taking care to previously cover all the very soft and swampy places with dried peat, sods, or brushwood: numerous drains are to be cut across the roadway from the one main drain to the other; they are to be three feet deep at the centre of the roadway, and four feet deep at the main drains: after the whole have remained in this state for two summer months, the bed for the roadway is to be neatly formed, with the sides on the same level, and with a convexity of half an inch in the yard.

"The carriage-way is then to be covered with six inches of clay, laid on evenly, and firmly compressed by stampers or rollers; it is to have a fall of one inch in the yard from the centre towards the sides: over the clay is to be put four inches of small gravel; it is to be frequently rolled, and, when solid and compressed, the foundation will be formed for the reception of the road materials."

CHAPTER IV.

DRAINAGE.

So much depends upon proper drainage, that too great attention cannot be given to this part of the business of road-making.

This operation should be carried on at the time of forming the road. When it is to be made over flat and wet land, open main drains should be cut on the field side of the fences: these drains should communicate with the natural watercourses of the adjacent land; their size should depend upon the nature of the country and upon local circumstances.

In general, these side drains should be cut at least three feet deep below the level of the bed of the road; they should be one foot wide at bottom, and five feet wide at top.

In crossing marshy land they should be made sufficiently deep and wide to obtain earth to raise the bed of the road, from side to side, three feet higher than the natural surface, in order to compress the subsoil and reduce its elasticity.

If, in consequence of the road running along the side of a hill, or passing through a cutting of a hill, or of the intervention of buildings or other obstructions, main open drains cannot be formed, it then becomes necessary to make covered drains on each side of the road. These should be formed of stone or brick, and should be strongly and substantially built. If built with stone, a flat stone should be laid at the bottom of the drain, the side walls should be not less than twelve inches thick, and built in regular level courses, but without mortar; they should be eighteen inches high, and twelve inches apart. Particular care must be taken that the covering stones have a bearing of at least four inches on the side walls. They should have a layer of brushwood put over them; and the drain should then be filled up with gravel, or small stones. In gravel countries, or where stone is difficult to be procured, it will be necessary to build the main side drains of brick; the side walls should be four inches thick, three bricks high, and five inches apart, and covered with brick on the flat: these covering bricks should not be laid close together; an interval of at least half an inch should be left between each, to allow the water to enter the drain from above.

In some cases it will be necessary to build circular brick drains twelve or eighteen inches in diameter, according to circumstances; but they are expensive, and require inlets, built with brick, with iron grates. In consequence of its being necessary to build these drains with mortar, they are not so good as the open-jointed drain last described, unless there is a considerable run of water. Plate II. fig. 12.

If springs rise in the site of the road, or in the slopes of deep cuttings, stone or tile drains should be made into them, so as completely to carry away all the water.

In cuttings it is necessary to make drains of small dimensions from the centre of the road to the side drains. These drains should form an angle in the centre of the road, in the shape of a V, technically called mitre drains: the angle or splay of these drains should depend upon the inclination of the road; it should not make the inclination of the drains exceed one inch in 100; for if it be greater, the run of the water will undermine the sides, and injure them. These mitre drains should be nine inches wide at bottom, twelve inches wide at top, and ten inches deep. These drains should be placed at about sixty yards from each other, or about thirty in the mile; but if the soil be wet, this number should be considerably increased. They are to be filled with rubble stone or cleansed gravel. If gravel is used, a draining tile should be laid along the bottom before the gravel is put on. In stiff and retentive clays these drains should be twelve inches wide at bottom, sixteen inches wide at top, and eighteen inches deep. In the bottom bricks should be laid on the flat, six inches apart; these should be covered by other bricks laid across, so as to form an open channel of six inches by four, and over these gravel or small stones should be laid. Where stones can be obtained, they will answer as well as bricks.

The upper part of these mitre drains should communicate with the road materials, so as to draw the water from them.

According to the inclinations of a road, and the form and wetness of the country through which it passes, cross drains of good masonry should be built beneath it, having their extremities carried under the road fences.

One of these drains should be made wherever the water would lie on one side of the road, and can only be got rid of by carrying it to the other side. When the road passes along the slope of a hill or mountain, a great number of these drains are necessary to carry off the water that collects in the channel on the side next the high ground. They should be placed at from 50 to 100 yards' distance from each other, according to the declivity of the hill; so that the side channels may not be cut by carrying water too far. In these situations inlets should be built of masonry, to carry the water from the side channel into the cross drains. The manner of building an inlet will be described in the chapter on Road Masonry. On flat ground numerous outlets should also be made from the side channels, under the footpaths, or wastes and fences, into the field ditches.

In mountainous countries, where the road passes along slopes, it is necessary to carry open or catch-water drains, branching from the upper ends of the cross drains, in an inclined direction, so as to catch the surface water before it can reach the road.

After all these precautions have been taken, the preservation of the surface from injury by water should be further secured, by giving a proper convexity in the cross section, and by making regular side channels.

These side channels will be formed by the angle where the curved surface of the road abuts on the footpath, or other defining boundary of the road-way. They will be capable of carrying off a great quantity of water, without being made in the form of a square-sided drain.

Attention in making the surface of a proper convex form is particularly necessary on hills, in order that the water may have a tendency to fall from the centre to the sides, in place of running from the sides to the middle of the road, which it certainly will do unless the side channels are kept below the centre, in the manner hereafter described.

On all hills the greatest care should also be taken to keep the side channels always open; for, if they are obstructed with weeds or mud, the water will find its way over the middle of the road. The side channels should be all thoroughly repaired as well as all the drains before the approach of winter, and again after the winter is over; but, besides these repairs at fixed periods, daily attention should be given to take care that no obstruction takes place.

Whenever a branch or field road joins a main road, it should not be allowed to interfere with the side channel: in order to secure this object, the point of junction should always be on the field side of the side channel; unless this is the case, the branch or field road will, when on a higher level than the main road, carry its surface water upon the main road.

In addition to all these means recommended to be adopted for securing the drainage of a road, it is of the utmost importance that evaporation should have full effect in drying up the surface, by allowing the sun and wind to act upon it in the freest manner.

The necessity of giving a road a good exposure has already been mentioned under the head of "Laying out a Road;" and the value of a rapid evaporation will be more fully explained when the repair of roads is brought under consideration.

Roads kept dry will be maintained in a good state with proportionally less expense. It has been well observed, that the statuary cannot saw his marble, nor the lapidary cut his jewels, without the assistance of the powder of the specific materials on which he is acting: this, when combined with water, produces sufficient attrition to accomplish his purpose.

A similar effect is produced on roads, since the reduced particles of the materials, when wet, assist the wheels in rapidly grinding down the surface.

A more particular description of the mode of constructing the several drains which have been mentioned will be given in the chapter on Road Masonry.

CHAPTER V.

DIFFERENT KINDS OF ROADS, AND MODES OF CONSTRUCTING THEM.

THERE is not to be found in any of the books on road-making a distinction between roads for great and roads for little traffic. Each author has written as if there ought to be only one kind of road for every kind of use. This is a great mistake, and has led to much confusion in forming opinions upon the proper construction of roads. A road of earth, put into a regular form, will answer for a park drive. The same with a coating of gravel will do for light carts and other carriages. So a road made with ten or twelve inches of broken stones laid on the natural soil, will be a sufficiently good road when the traffic is not considerable; while very great traffic requires a road to be constructed in a different manner, that is, with much greater solidity and hardness, so as to allow carriages to be drawn on it with the smallest possible quantity of tractive power, the object that all road-makers ought to have in view.

The different kinds of roads may be distinguished and described as follows:—

1st. Iron railways.

2d. Paved roads.

3d. Roads of which the surface is partly paved and partly made with broken stones or other materials.

4th. Roads with a foundation of pavement and a surface of broken stones.

5th. Roads with a foundation of rubble stones, and a surface of broken stones or gravel.

6th. Roads made with broken stones laid on the natural soil.

7th. Roads made with gravel laid on the natural soil.

IRON RAILWAYS.

The uses and advantages of iron railways with locomotive engines have of late been so fully explained in several works of great ability that it is not necessary to repeat in this what was stated respecting them in the last edition.

The eagerness which was so generally displayed by vast numbers of persons to give credit to the representations of the great profits to be realized by railway shares, gave so much encouragement to all those adventurers, who looked to derive immediate advantage from railway projects, that acts of parliament have been passed for railways in every part of the kingdom. The experience, however, which has been gained from those already completed, and from the enormous expense incurred on those which are in progress, has led to a general opinion that there is little probability of more than a few of these works affording any ultimate return for the money expended upon them.

The heavy expense which is proved by experience to be unavoidable in keeping the railways and engines in repair, where great speed is the object, will in numerous cases soon make it evident that no dividends can be paid to the shareholders; and the cheaper method of using horse power will be adopted. This has recently happened on a railway, where, although the traffic on it was very considerable both of goods and passengers, the cost of using steam power absorded nearly all the money received; and accordingly, a case having been made out by an eminent engineer to show that if horse power were employed the traffic would afford a dividend, the use of steam power was discontinued, and the result has proved the change to be completely successful.

What seems to have been the great error on the part of those who have introduced the modern railway system was making excessive speed the main object of it. It is this which has led to the enormous expense, 1st, as to the gradations of the lines; 2dly, as to the strength of the construction of railways; and 3dly, as to the engine. But the attaining of the speed of 25 or 30 miles an hour, at such an enormous expense, cannot be justified on any principle of national utility. The usefulness of communication, in a national point of view, consists principally in rendering the conveyance of all the productions of the soil and of industry as cheap as possible. This keeps down the prices of food, the prices of raw materials, the prices of finished goods; and thus increases the consumption of all productions, the employment of labour and capital, and generally the national industry and national wealth. But a speed of 10 miles an hour would have accomplished all these purposes, and have been of great benefit to travellers, while it could have been attained at from one half to one third of the expense which has been incurred by the system that has been acted upon. It is no doubt true that travelling at the rate of 25 or 30 miles an hour is personally very convenient, but how it can be made to act so as to contribute very much to the benefit of the country at large it is not easy to discover. Economy of time in an industrious country is unquestionably of immense importance, but after the means of moving at the rate of ten miles an hour is universally established, there seems to be no very great advantage to be derived from going faster.

The use of steam power and the practice of keeping up an excessive rate of speed has necessarily led to high charges for carrying passengers and goods. A slower rate of speed would, by diminishing expense, admit of the charges being moderate, and in this way the national interests would be best promoted. The object in making railways ought, from the beginning, to have been the reduction of the cost of moving passengers and goods to the lowest possible limit, and not excessive speed. This would have made the money applied to railways go much farther in extending them over the face of the country; the risks of accidents would have been almost wholly avoided; while the charges for travelling and transporting goods would have been considerably less. It is, however, right to admit, that if the raging passion for excessive speed had not been gratified, subscribers, probably, would not have been found for forming railway companies, and what was really useful and necessary in substituting railways for common roads would never have been accomplished. The public, in fact, are alone to blame for the immense waste of money which has taken place in forcing an excessive rate of speed, and in producing that superfluity of embellishment and grandeur which is to be seen on all the railways.

PAVED ROADS.

In situations where canals cannot be constructed, either for want of water or other circumstances, and where the description and quantity of traffic, or local obstructions, do not justify the expense of forming a railway, paved trackways and roads made on proper principles would be found much better for conveying goods than turnpike roads, constructed as they usually are. The advantages which may be derived from paved roads as a means of transport have been too much overlooked; and therefore it is very important to show how much superior a well-made pavement is to a common road in enabling horses to draw very large burdens.

The plan of paving which contributes most to diminish the labour of moving heavy weights on roads, is that of forming as hard and smooth a surface as can be formed, with stone for the wheels of carriages to roll upon. This is effected by making use of large blocks of granite or other hard stone. Roads of this kind, when the blocks are about 16 inches wide, and are laid in parallel lines, are commonly called stone tramways or trackways.

On a well-constructed road or trackway of this kind, it has been proved by experiment that a London draught horse can draw, on the level, ten tons.*

In cases where hills cannot be brought to a better inclination than 1 in 20 without incurring a very heavy expense, a stone trackway may be used with very great benefit. On the Holyhead Road, between Towcester and Daventry, there is a hill one mile in length, from Forster's Booth to Geese Bridge, and another hill from this bridge, of the same length, to Stow. The inclinations on both hills being about 1 in 20, the estimated expense for reducing these inclinations to 1 in 24 was 20,000*l.*; but nearly all the advantage in diminishing tractive force which could have been obtained by this outlay has been obtained by a moderate extent of cutting and embanking, and by making a stone trackway on a portion of each hill; the whole at an expense of 9,000*l.* The power required to draw one ton on the old road over those parts of the hills which were at 1 in 20, was found to be by Mr. Macneill's instrument 294 lbs., but the power required on

* See Report of Mr. Walker on the Commercial Road, and Remarks of Mr. Macneill on Tramways, Appendix, No. III.

these trackways for the same weight and at the same rate of inclination is only 132 lbs.; so that the tractive force has been reduced more than one half by this improvement.

The Commercial Road, from the West India Docks to Whitechapel, was made on this plan in 1829, that is, with large blocks of granite five or six feet in length, sixteen inches wide, and twelve inches deep, laid for the wheels to run upon, as on a tramroad of iron, except that there is no flanche. The space between the granite blocks is paved. The plan has succeeded, as may be seen from the following Report of Mr. Walker to the trustees of this road: —

"I beg to report the results of the experiments made this day upon the stone tramway now forming on the Commercial Road, before you, accompanied by the Chairman of the West India Dock Company, and Mr. Colville, one of the directors.

"The experiments were made upon the space between the West India Dock-gate and the first turnpike upon the Commercial Road, with a very good town-made waggon, belonging to Messrs. Smith and Sons, distillers, and a stone truck, belonging to Messrs. Freeman.

"The dust had been swept off the tramway in the morning. The distance is 550 feet, of which 250 feet nearest the Dock-gate rise 1 foot, or 1 in 250, and the other 300 feet rise about 2½ feet, or 1 in 116.

"The whole rise in the 550 feet is 3½ feet, or 1 in 155.

"The gravity of one ton upon the lower length is, therefore, 2,240 lbs. divided by 250, or nearly 9 lbs. Upon the upper length it is 2,240 lbs. divided by 116, or 19½ lbs., and the average of gravity upon the whole length is 2,240 divided by 155, or 14½ lbs.

"Experiment 1st. The general average resistance of four tons gross (viz. waggon 1 ton 16 cwt. and goods 2 tons 4 cwt.), as ascertained by your chairman (C. H. Turner, Esq.) and Mr. Colville, by means of a spring weighing machine, was 127 lbs.; from which, if we deduct the gravity of 4 tons, or 19½ lbs. multiplied by 4, say 77 lbs., there is left, for the friction of 4 tons, 50 lbs., which gives for the friction of 1 ton 12½lbs., $\frac{1}{180}$th of the whole weight moved.

"This friction is not more than upon the best constructed edge railway. I consider that the greater size of our wheels, and there being no flanche, compensate for the roughness of the stones (from their being newly laid,) as compared with an iron railway.*

"Experiment 2d. A pony 12½ hands high, weight 4½ cwt., drew upon the upper part in your presence, and afterwards upon the lower part in your and the directors' presence, six tons (gross). I was not aware that the difference of inclination of the two parts was so great, or he should have

* In consequence of the improved mode of laying rails, the resistance from friction is now about 8 lbs.

gone over the upper length again, — he had done it more than once before.

"Taking, therefore, the upper part on the rise of 1 in 116, the pony's exertion was,

	lbs.
Gravity $19\frac{1}{3}$ lbs. multiplied by 6, or -	116
Friction $12\frac{1}{2}$ lbs. multiplied by 6, or -	75
Making together	191

and 191 lbs. divided by $12\frac{1}{2}$ lbs. (the friction of one ton) gives 15 tons.

"The pony's work, therefore, was equal to *fifteen tons* drawn upon a level road.

"Experiment 3d. The waggon, loaded as in the preceding experiment, being turned round and started by the pony's exertion, ran down the whole length to the Dock-gate with increasing velocity (the pony not drawing it), and for a distance off the tramway, before it could be stopped; consequently the average fall of 1 in 155 exceeded the resistance by friction.

"Experiment 4th. A powerful horse (weight 14 cwt.) drew 12 tons gross (the waggon and truck loaded) from the West India Dock-gate to the turnpike at the rate of 4 miles per hour.

"Taking then the upper length, or a rise of 1 in 116, we have,

	lbs.
Gravity 12 times $19\frac{1}{3}$ lbs., or -	232
Friction 12 times $12\frac{1}{2}$ lbs., or -	150
Making together	382

and 382 lbs. divided by $12\frac{1}{2}$ lbs. gives $30\frac{1}{2}$ tons.

"The horse's work was therefore the same as if he had been drawing 30½ tons upon the level.

"The full average work of a horse per day is 150 lbs. moved 20 miles; consequently the pony was exerting one fourth more than the average work of one horse through the day; and the horse was doing the work of 2½ horses.

"The horse appeared to go easily; but the exertion was, of course, too great to be continued for any considerable time, so as to form a basis for general calculation.

"Upon the whole, I think the conclusion is, that if the road were level, the work of a London draught horse upon the tramway would be ten tons (gross); but as the Commercial Road rises towards London, a deduction must be made from this for gravity, the amount of which depends upon the inclination of the road, and is common to all kinds of roads and railways. Therefore, taking all things into consideration, I am of opinion that six tons (gross) from the Docks to Whitechapel, and a greater weight from Whitechapel to the Docks, may be considered a proper load for one horse on the tramway." *

* The following is the specification for making a road with granite blocks on each side of the Geese Bridge Valley, on the Holyhead Road:—

"On the steep parts of the road, as shown on the section from the letter A to the letter B (on the east side of the valley, and from C to D on the west side), two lines of stone blocks are to be laid down, to form a tram or track for the wheels of carriages. These blocks are to be of granite, from the quarries at Mount Sorrel. They are to be twelve inches deep, fourteen

In proportion as roads or streets are paved with proper skill and care, the advantages in diminishing

---

inches broad, and not less than four feet long. They are to be pricked on each face, the upper surface to be parallel to the bed, and the sides and ends dressed square, and pricked in the same manner as the specimen now lying at Snow Hill Wharf, near Weedon, for the inspection of those who may propose for the work. The foundation for these blocks is to be prepared in the following manner, as shown in plan 3.:—The ground is to be excavated to the proper depth, to receive the different materials hereafter described, so that the upper surface of the stone blocks will be on a level with the carriage-way when finished. The width of this opening is to be eight feet, extending from the channel towards the centre of the road. When the bottom is properly and evenly levelled, a pavement of limestone, or good sandstone, from the quarries near Stow, is to be laid by hand in a similar manner to that described for the road-work. These stones are to be eight inches deep throughout, to be firmly packed, and perfectly level from side to side. When the interstices are filled up with stone chips, and the whole firmly packed, good lime grout is to be poured into the joints. One barrel of lime and two barrels of sand and gravel, properly mixed, and used whilst fresh, are to be used on every lineal yard. This work is to be done to the satisfaction of the engineer, and every care taken to render it solid and substantial in every respect. On this pavement three inches of broken stone from Lovel's or Stow quarries are to be laid: no stone is to exceed one inch and a half in its largest dimensions. They are to be put on after the foundation course has been inspected and approved by the engineer, and not before. On this layer of stone two inches of the best gravel is to be laid: it is to be rolled with a metal roller several times, so as to consolidate the stone and gravel together, and render it one firm mass. On this gravel the stone blocks or trams are to be laid: they must be even on the bed, and the ends must abut together, and form one continuous and uniform line. The surface of each must be on the same level, and perfectly uniform with the blocks the tractive force required for drawing carriages upon them will be augmented. On a smooth,

on the opposite side; that is to say, the surface of each line must be perfectly level, and correspond throughout. When the trams or blocks are laid to the satisfaction of the engineer, the space below them and the space on the outside is to be filled up to within six inches of the upper surface with broken limestone, no piece of which shall exceed one inch and a half in its largest dimensions. A row of paving-stones of granite, not less than six inches deep, five inches wide, and nine inches long, is to be placed on each side of the trams or blocks: these stones to be of the best granite from Mount Sorrel quarries. They are to be dressed in the same manner as the stones called sovereigns in the London market. These paving-stones are to be placed close to each side of the blocks, and firmly packed up with Hartshill stone, broken not to exceed two inches in its largest dimensions for the first three inches; the next three inches towards the surface to be broken to one inch and a half in its largest dimensions. When this work is completed, the whole is to have a top dressing of one inch of good gravel, evenly spread, and rolled twice in wet weather. The sides of the road next the trams is to be formed and dressed so as to come on the same level, and when the whole is finished to form one uniform surface of carriage-way, as shown on the plan.

"A drain, eight inches deep and four inches wide, is to be formed on each side of the trams, the bottom of which is to be eight inches below the bottom course of the pavement, and is to be formed of rubble-stone or bricks, so as to lead off any under or surface water that might otherwise injure the foundation or bed of the trams. These drains are to have off-lets into the large side drains, which are to be two feet under the surface of the road-way where the trams are to be used; and, should it be thought necessary, wells or upright shafts must be constructed at every 100 yards apart through the cuttings, to let the surface water into the under drains, without running along the surface. The tops of these wells are to have an iron grate one foot square, the bars of which are to be two inches deep, one inch and a half apart, and one inch wide, set in proper frames." —See Appendix, No. III. on Tramways.

well-made pavement, quite horizontal, it appears, from the experiments made with Mr. Macneill's machine, that the resistance to draught is not more than the 100th part of the weight of the carriage and its load, when the carriage is properly constructed, and mounted on straight and cylindrical axles.* According to this, a horse of great power would be able to draw on such a road, if horizontal, six tons and three quarters; and if with no greater inclination than 1 in 50, two tons and a quarter.

The same experiments show that the tractive force on a good pavement is 33 lbs., while that on a broken stone road is 65 lbs., and that on a gravel road 147 lbs.

The following statement on pavements is taken from the evidence of Mr. Walker, given before a select committee of the House of Commons, upon the Commercial Road from London to the West India Docks:—

"It is not, I am sure, overstating the advantages of paving, but rather otherwise, to say, taking the year through, two horses will do as much or more work with the same labour to themselves upon a paved road than three upon a gravelled road, if the traffic on the gravelled road is considerable; and if the effect of this is brought into figures, the saving of the expense of carriage will be found to be very great when compared with the cost of paving. If the annual tonnage upon the Commercial Road be taken at 250,000 tons, and at the

* See Mr. Telford's Seventh Report on the Holyhead Road.

rate of only three shillings per ton from the Docks, it could not, on a gravel road, be done under four shillings and six-pence; say, however, four shillings, or one third per ton difference, which makes a saving of 12,500*l.* in one year.

"I think I am under the mark in all these figures; and I am convinced, therefore, that the introduction of paving would, in many cases, be productive of great advantages."

Mr. Walker further says, that "during thirteen years that the East India Docks' branch has been paved, the paving has not cost twenty pounds in repairs, although the waggons, each weighing about five tons, with the whole of the East India produce, which is brought from the Docks by land, have passed all that time in one track upon it; and a great deal of heavy country traffic for the last eight years, when a communication was formed with the county of Essex."*

This road has been referred to, not by way of showing a perfect specimen of pavement, but generally to point out the advantages of paved roads; for this road, in consequence of the plan of Mr. Walker not having been strictly attended to, was by no means originally constructed in so perfect a manner as it might have been.

Still, however, it was by far the best specimen of pavement that had been executed; and it has fully established, by experience, the great advan-

* See Report of the Committee of the House of Commons on Turnpike Roads in 1819.

tages which the public may obtain by making a paved road for transporting merchandise, when it is not possible to make a canal or railway.

A common opinion prevails, that because paved streets have almost every where been suffered to be rough and imperfect, all pavements must necessarily be rough and bad; but a slight degree of consideration will show that this opinion is without foundation, and that, in point of fact, the cause of rough and bad pavements is bad management, arising from the ignorance of those employed to make them, or the want of sufficient funds for executing good work.

The chief defects of all pavements arise from neglecting to give the stones a proper shape, and to construct a substantial foundation to support them.

The foundation for the London pavements has commonly been formed with all sorts of old rotten materials, without having uniformity in texture or solidity.

This makes the bottoming of unequal strength, and the consequence is, that after the pavement has been laid upon it, the weaker parts give way, while the stronger remain firm, so that the surface of the street becomes low in some places, and high in others, and very soon rough and out of order.

But the effect of this bottoming is visible in another way; for in consequence of the defective shape of the stones, and also of the defective manner in which they are set, there is a quantity of soft earth between them, which serves to conduct the water which falls on the pavement through the joints of the stones into the body of earth that lies under them. But when the water gets into this earth, it forms a bed of mud, and the heavy weights passing over the paving-stones press them into it, and then the mud, not being able to resist the pressure, rises upwards between the joints to the surface of the pavement.

In making some of the new pavements in London, more attention has been paid to the foundation than formerly, but still not so much as there should have been. Fleet Street, for instance, was paved with considerable care; the stones were properly shaped and dressed, and evenly laid: the ground was dug out twelve or eighteen inches deep, and a body of broken stones was put into the space thus cleared out for a foundation, and the joints of the paving-stones, when set, were grouted with liquid lime. Yet, notwithstanding all this, the pavement soon got out of order, and became uneven and extremely defective in a few months after it was made. The cause of this failure was that the stones for the foundation were thrown in loosely, by cart-loads at a time, and merely levelled before the paving-stones were laid on.

The great defects in the London pavement, with respect to the shape of the paving-stones, and to proper bottoming, are, in a great measure, to be attributed to the errors committed by the persons who have had the management of making contracts for it.

They have acted too much on the principle of

getting cheap work, and to accomplish this have neglected the main point of securing good work. In this way they have promoted a system of inconsiderate competition, and thereby reduced the price of paving work so unreasonably as to make it impossible for contractors to provide the best materials, and to bestow the necessary portion of labour on dressing and setting them, without loss.

Besides this, the managers of the pavements have sometimes committed another error, in requiring the contractors to execute their work by the superficial square yard, of a certain depth, without adding conditions respecting the weight of the stones to be put into a square yard, and without requiring the close joining of the stones. As contractors purchase the stones by weight, the less square (that is, the more imperfect) they are, the more profit they will have on each superficial yard; so also in proportion as the joints between the stones are wide, fewer stones will be used, and the work will be proportionably profitable to them, and defective for the use of the public.

Having thus shortly explained the principal defects of the London practice of paving, the mode that seems fit to be adopted in its stead shall now be described.

The first object to be secured is a good foundation. In those streets where there is a constant passing of carriages, both night and day, such as the Strand, Fleet Street, Holborn, Cheapside, Piccadilly, and Oxford Street, the foundation for the pavement should be a sub-pavement made of old paving stones, or any kind of coarse stone; and this should be laid on a bed of broken stones. This mode of paving is in use in Paris. In the autumn of 1835, for instance, the old pavement of the Rue Dauphine was taken up, and relaid on a bed of gravel, to form the foundation for the new pavement. This practice has proved completely successful. In streets where the traffic is not very great, the foundation should be made in the following manner:—a bed should be formed with a convexity of two inches to ten feet, so as to admit of twelve inches of broken stones being laid upon it. These should be put on in layers of four inches at a time. After the first layer is put on, the street should be kept open for carriages to pass over it. When this first layer has become firm and consolidated, then another layer of four inches should be put on, and worked in as before, care being taken to rake the ruts and tracks of the wheels of carriages, so that the surface may become smooth, and consolidated. The same process should be repeated with the third layer of stones, by which means a solid and firm foundation will be established, of twelve inches in thickness, for the dressed paving stones to lie upon.*

* Mr. Edgeworth says, in his Essay on Roads, " In all pavements, the first thing to be attended to is the foundation. This must be made of strong and uniform materials, well rammed together, and accurately formed to correspond with the figure of the superincumbent pavement. This has nowhere been more effectually accomplished than in some lately made pavement in Dublin. Major Taylor, who is at the head of the paving board,

After this operation is accomplished, the next thing to be attended to is to provide proper paving-stones. These should be cut into a rectangular shape, and of the hardest quality that can be procured: granite is the best, but whinstone, some descriptions of limestone, and freestone, will answer the purpose.

With regard to the size of the stones, that should be regulated by the intercourse.* The streets should be divided into three classes, according as the thoroughfare is greater or less. For streets of the first class, or greatest thoroughfare, the stones should be ten inches in depth, from ten to fifteen inches in length, and from six to eight inches in breadth on the face. For streets of the second class, the stones should be eight inches in depth, from eight to twelve inches in length, and from five to seven inches in breadth on the face. For streets of the third class, the stones should be six inches in depth, from six to ten inches in length, and from four to six inches in breadth on the face.

After having prepared a proper bottoming, the greatest care must then be bestowed in setting the stones. Fine gravel must be provided, cleansed

---

before he began to pave a street, first made a good gravel road, and left it to be beaten down by carriages for several months: it then became a fit foundation for a good pavement." Page 33.

* "In France, pavements are made with stones cut with the hammer, of seven or eight inches on every side, like so many dice." — *Encyclopédie de l'Ingénieur*, vol. i. p. 355.

from all earth, to form a bed over the bottoming of two inches thick for the stones to be set in. Strong mortar must also be provided; and, besides the common tools, each pavior should have a wooden maul, the head of which should be made of beech or elm, and should weigh about fourteen pounds. The stones should be selected so that they may be laid in even courses, and so as to match, as nearly as possible, in each course with regard to breadth and depth.

The pavior should first set a stone on the gravel bed, by striking it strongly downwards with the maul, and then on its sides. Then he should lift it out of its berth, and put mortar on the sides of the two adjoining stones; after which he should again place the stone in its berth, and strike it as hard as he can, downwards and sideways, with the maul, till it is fastended in the position in which it is to remain. Each stone should be set in this manner; and, when the pavement is finished, it will be so firm as not to require ramming.*

The crossings for foot-passengers should be raised above the level of the pavement, by giving a moderate convexity to the bottoming. They should be made with stones of the size for streets of the first class, more accurately dressed.

The pavement should be formed with a regular, but very moderate, convex surface, by giving the bed for it the convexity already mentioned: there

* See Mr. Telford's Report on Pavements, in the Appendix, No. 2., and Mr. James Walker's observations on the same subject, Appendix, No. 3.

should be no gutter or other channel but that which will be formed by the angle made by the surface of the pavement abutting on the curb-stone. The curb-stone of the pavement should be made of long blocks of stone, of a quality sufficiently hard to resist the shocks of wheels striking it. These blocks should be bedded in gravel, and joined with cement: they should be sunk four inches at least into the ground, and should be six inches above the pavement.

The foot pavements should be made of well-dressed flags; each flag to have its sides rectangular, and to be set in mortar with a very close joint upon a strong gravel bed of six inches in depth. The flagstones should be at least two inches and a half thick: the surface of the foot pavements should have a declivity at the rate of one inch in ten feet towards the street.

In those towns where the intercourse of foot-passengers is considerable, the greatest possible breadth should be given to the foot pavements. In general, this important accommodation is too much neglected.

In making a contract for paving a street, it should be so arranged that the work shall be paid for by the superficial yard. A specification should be connected with the contract, requiring that the natural soil shall be removed to a certain depth, and made level; that the bottoming shall consist of certain prescribed materials, and of a fixed depth and convexity, and be laid on in the manner already described; that the stones shall be of a certain shape, and size, and weight, in each superficial yard; and that they shall be set in a regular manner.

The specification should also provide that an inspector shall be appointed to see that the conditions of the contract are fulfilled.

A drawing of the transverse section of the proposed pavement should be made, and attached to the contract. (See Plate III., Fig. 1.) But, besides these precautions, the persons who have the management of the business of making the contract should take care to ascertain the sum for which an honest and skilful contractor can execute the work in a proper manner. This is necessary in order that they may avoid making a bargain on such terms as will oblige the contractor to have recourse to inferior materials and workmanship to save himself from losing money by his undertaking. It is not, by any means, difficult to ascertain exactly what each part of this kind of work will cost, and to make an accurate estimate of the total expense to be incurred.

Such an estimate ought always to be made by a competent person in the employment of the commissioners or other persons who have to manage contracts for pavements. And in place of promoting an indiscriminate competition for the purpose of lowering the price, without regard to the quality of the work to be performed, the commissioners or other managers ought to call upon a few of the most respectable paving contractors for proposals; and give the contract to the bidder who comes nearest to the estimate of their own officer. If no fair tender be made, recourse may then be had to advertising for a contract.

With respect to repairing paved streets, this work should be done by contract. The price may be fixed by the superficial yard; but the manner of doing the work should be minutely described in a specification.

Whenever any pavement is taken up, if only a few stones, there ought to be fresh broken stones provided for making good the bottoming. A principal cause of the bad state of the pavement in London is the neglect of timely repair. After a pavement is newly laid it is usually left without any repairs until it is in a ruinous state; but, instead of this, constant attention should be paid to it, and as soon as a single stone gets out of its proper bearing it should be taken up, and relaid with new bottoming. In case of breaking up the pavement for water or gas pipes, it should be specified that a complete bottoming of stones should be first laid down over the pipes, similar in every respect to that provided for the first making of the pavement.

The paving-stones should be laid on loose at first, and left till the bottoming is consolidated, and then they should be taken up and carefully set in mortar.

Paved streets have been objected to on account of the noise made by carriages passing over them. The noise chiefly arises from the boxes of the wheels striking the arms of the axletrees; and, therefore, when a paved street is exceedingly rough, the strokes of the axles are frequent and violent. But when a paved street is properly made, the surface of it will be comparatively smooth, and then both the number and force of the strokes of the axles on the boxes, and consequently the noise made by carriages, will be greatly reduced. When a carriage passes from a rough to a well-made pavement, the difference of sound is immediately perceivable.

It is supposed by some persons that if the streets were paved in the way proposed their surface would be too smooth for horses to go safely over them; but this supposition is not well founded, except when that kind of stone is used which becomes polished by wear.

Scotch granite and some other kinds of stone do not become polished; and therefore pavements made with them will never have so smooth a surface as to be unfit for horses. A horse properly shod will seldom slip on a pavement, or fall, unless when thrown down by being turned too short, or other careless management.

The enormous expense which has been incurred by adopting the plan of broken stone streets in London, in place of pavements, is fully established by the following return, which was presented to the House of Commons in the year 1827.

By this return it appears that the first cost of converting 1 mile 250 yards from a pavement into broken stone road was 12,842*l.*, and that the annual expense of maintaining this 1 mile 250 yards has been 4,003*l.*, being at the rate of 1*s.* 9*d.* per superficial square yard.

"M'ADAM'S ROADS:

"*Regent Street, Whitehall, and Palace Yard Streets.*

"Account of all sums expended by the commissioners acting under 5 Geo. IV. c. 100. and 6 Geo. IV. c. 38., in converting Regent Street, Whitehall, and Palace Yard Streets into broken stone roads, including the value of the paving-stones converted into broken stones; also, of the expense incurred in maintaining these roads in repair, including scraping and watering, and all other expenses in the year ending 5th January 1827; and showing the number of lineal yards, and of superficial square yards, in the Regent Street District, the Whitehall District, and Palace Yard District; viz.

	£	*s.*	*d.*
"The cost of converting Regent Street, Whitehall, and Palace Yard Streets into broken stone roads has been - - -	6,055	8	3
"And the value of the old pavement taken up and broken for that purpose is estimated at -	6,787	7	0
	£12,842	15	3

	£	s.	d.
"The cost of maintaining these roads in repair, including scraping, and every expense, except watering, in the year ended 5th January 1827, was - - - -	4,003	18	4
"The cost of watering the said roads in the year ended 5th January 1827, was - - - -	628	11	0
	£4,632	9	4

"The extent of the said roads is as under; viz.

	Yards Lineal.	Yards Superficial.
Regent Street and Waterloo Place, from Oxford Street to Pall Mall - -	1,300	24,401
Whitehall, from Buckingham Court to Richmond Terrace - - -	450	11,651
Margaret Street and Old and New Palace Yards - - -	260	9,199
	2,010	45,251

"(Signed) A. M. Robertson,

"Clerk to the Commissioners for carrying into execution the Acts 5 Geo. IV. c. 100. and 6 Geo. IV. c. 38.

"Office of Woods, &c.
30th April 1827."

The following is a copy of a statement which appeared in a London Morning Paper on this subject:—

"*Macadamisation.*

"In the proceedings taken before the House of Lords on the 11th of last May, several witnesses gave evidence on the Westminster Improvement Bill as to the comparative expense of macadamising and paving. According to this evidence, there is no less a difference in ten years than 2*l.* on every superficial yard; a yard of paving for that time amounting to 10*s.* 10*d.*, and a yard of macadamised road for the same period costing 2*l.* 10*s.* 10*d.*

"Mr. Johnson, an eminent pavior and stone merchant, stated before their Lordships that he had been a contractor for St. George's, St. Ann's, St. Giles's, and other parishes, and for some parts of the city, which enabled him to make very accurate calculations. He proved that the very best pavement would cost 13*s.* per square yard, which would require no repair for the first year certainly, and in most cases would cost nothing in repair for the first three years; that the expense after the first year would be about 4*d.* per yard per annum for ten years, after which the pavement as laid down would be worth 8*s.* per yard to the parish; thereby reducing the expense of a square yard of pavement in ten years to 10*s.* 10*d.*, as under:—

	s.	d.
First cost, per superficial yard -	13	0
Ten years' repairs, at 4*d*. - -	3	4
Ditto cleansing, at 3*d*. - -	2	6
	18	10
Deduct value of old stone - -	8	0
Per yard -	10	10

"The old stone might last twenty years longer; but, at all events, would be worth 8*s*. per yard after ten years wear. That statement was made on a calculation of using the very best material; but most of the pavement is laid down at from 7*s*. to 10*s*. per yard.

"A macadamised, or broken stone road, requires for keeping in repair the first year, and every year after, two coats of three inches thick, to allow for wear: the coating costs 1*s*. 9*d*. each yard: the cleansing and scraping cost 10*d*. each yard, as under:—

	£	s.	d.
First cost, per superficial yard - -	0	7	6
Two coatings, at 1*s*. 9*d*. each per yard, for ten years - - - -	1	15	0
Cleansing, at 10*d*. per yard, for ten years - - - - -	0	8	4
Per yard -	2	10	10

"The surveyor to the Commissioners of Westminster Bridge stated that the expense of paving and keeping in repair the bridge for twenty-two years (from 1802 to 1824) was 3,494*l.*, including 1,165*l.* for new pavement in the first year, making an annual expense of 159*l.* About two years ago the bridge was macadamised, and the year after cost 1,507*l.* 12*s.* 6*d.* There was a covering ordered in June 1825, which cost 172*l.* 10*s.*, besides Mr. M'Adam's annual charge of 300*l.* The surveyor said he thought it now required another covering like that of last year, at the expense of 470*l.* 10*s.*, as he had examined the road, and found the broken stones, on an average, not more than three inches thick."

The following is also taken from a London paper, and shows what was the result of converting the pavement over Blackfriars' Bridge into a broken stone roadway:—

Blackfriars' Bridge.

" The report presented to the Court of Common Council last week, from the General Purpose Committee, relative to Blackfriars' Bridge, stated, that the City Surveyor having declared Mr. M'Adam had completed his contract for macadamising the same, the Committee had subsequently employed him to keep the bridge in repair, and that he had since delivered in a bill to them for no less than 473*l.* odd, for such repairs, during a period of only eighteen weeks. The Committee further stated, that they had advertised for tenders to keep the said bridge in repair for twelve months, and several offers had been made them; one offering to do the same at between 300*l.* and 400*l.*, while a second tender was so high as 900*l.* In fact, it appears that the traffic over this bridge, which has greatly increased since it has been macadamised, is now to that amazing extent, that the new granite is ground to powder almost as quickly as it is laid down. It being thus evident that, to keep it in a proper state, the bridge would cost 1,000*l.* per annum, and the City having no separate funds for that purpose, the Committee recommended that it should be repaved on its present surface, on an estimated cost of 1,500*l.*; the expense of keeping which in repair used to average under 120*l.* per annum. The report further stated that Mr. M'Adam offered, at the time of the alteration being effected, to keep the bridge in repair for 130*l.* per annum."

In streets where the traffic is very great, a great saving of expense might be effected by paving about fifteen or twenty feet on each side, and having only about twenty or twenty-four feet of broken stone road in the middle. The drivers of heavily-laden waggons and carts would walk with more security on the sides than in the middle; and they could, therefore, drive their waggons and carts on the pavement, while the middle would be left free for lighter carriages.

In all hot climates roads should be paved, because neither small broken stones nor gravel will become bound together in a solid mass without moisture.

ROADS PARTLY PAVED AND PARTLY MADE WITH BROKEN STONES.

To make a road of this kind, sixteen feet of the middle should be perfectly well paved, according to the rules already laid down, and the remainder of it on each side of the pavement should be made with small broken stones. The advantages that would be derived from such a road would be—1st, to save the labour of horses, as before explained in treating of pavements; and, 2dly, to diminish the expense of repairs.

If a road of this description be constructed with good materials, and in a workmanlike manner, it will require but a moderate expense to maintain it in excellent order; but constant attention will be necessary to keep the part where the pavement and the broken stones join from being cut into ruts.

Whenever the traffic of a road is so great as to wear down three inches of hard broken stones in a year, the middle part of it should be paved. At this rate of wear half a cubic yard of materials will be requisite for every lineal yard of eighteen feet of the breadth of the road. This will make the expense of new stones alone, for a road thirty-six feet wide, per mile per annum, (supposing the cubic yard of broken stones to cost twelve shillings,) amount to 1,056*l.*

As it has been already shown that while the tractive force on a good pavement is 33 lbs., that upon a gravel road is 147 lbs., or, in other words, that the labour of horses in drawing over gravel roads is more than four times as great as it is in drawing over pavements, it ought to be an invariable rule, in those districts where no other materials can be procured except gravel, to use pavements for those roads on which there is a large traffic. Thus, for instance, all the principal roads from London to the stone districts ought to be paved. If the pavements were perfectly well made, and all the hills reduced to inclinations of 1 in 40, a single horse would be able, without difficulty, to draw two tons weight upon them, and the present charges for the carriage of goods and merchandise might be reduced one half.

The following remarks of Mr. Telford on roads paved in the middle are taken from his first annual report on the Holyhead Road, dated 6th May 1824, p. 7 :—

"As there is an ascent the whole way from London to the Archway road, it is particularly desirable to have the road surface as hard as possible. Flints are much too weak. What has been done for improving the Kensington road suggests a proper remedy, and the complete success of this experiment fully justifies the same plan being adopted on this trust.* I am fully aware of the

* At the time when this was written a considerable portion of the old pavement remained. It had been laid down upwards of thirty years, had cost very little in repairs; and the complete success of the experiment, on a road formerly so bad, and on one of such great traffic, was fully acknowledged by all scientific road-makers. But as soon as the Metropolis Road Commissioners

strong prejudice against paved roads; but these prejudices have been created by the total want of skill in paving the streets of London, and their present very imperfect state. But if the middle sixteen feet in breadth of the present road were made use of as a foundation for a well-constructed pavement of stones of a moderate size, and of a cubical shape, with square joints, no road would be fitter for heavily-loaded carriages: the rough faces of the stones give a degree of action to the springs that eases the draft, while the perfect hardness permits carriages to move forward with a slight exertion by the horses. Along each side of the middle part there may be a roadway, twelve feet in width, made with broken stones of a durable quality. These side roads would answer for the lighter description of carriages and riding-horses. The whole breadth of the road, exclusive of footpaths, would, according to this plan, be forty feet. This appears to combine all the requisites for roads in the vicinity of a great city." (See Plate III. Fig. 2.)

A ROAD WITH A FOUNDATION OF PAVEMENT AND A SURFACE OF BROKEN STONES.

The following specification of the manner of constructing a road of this kind of thirty feet in

got the management of this road, they began to remove the pavement, and to substitute broken stones for it; and thus, without being conscious of it, they have destroyed the most perfect piece of turnpike road in England.

width, is taken from a contract for making a part of the Holyhead Road:—

"Upon the level bed prepared for the road materials, a bottom course or layer of stones is to be set by hand in form of a close firm pavement: the stones set in the middle of the road are to be seven inches in depth; at nine feet from the centre, five inches; at twelve from the centre, four inches; and at fifteen feet, three inches. They are to be set on their broadest edges lengthwise across the road, and the breadth of the upper edge is not to exceed four inches in any case. All the irregularities of the upper part of the said pavement are to be broken off by the hammer, and all the interstices to be filled with stone chips firmly wedged or packed by hand with a light hammer; so that when the whole pavement is finished, there shall be a convexity of four inches in the breadth of fifteen feet from the centre.

"The middle eighteen feet of pavement is to be coated with hard stones to the depth of six inches. Four of these six inches to be first put on, and worked in by carriages and horses; care being taken to rake in the ruts until the surface becomes firm and consolidated, after which the remaining two inches are to be put on.

"The whole of this stone is to be broken into pieces as nearly cubical as possible, so that the largest piece, in its longest dimensions, may pass through a ring of two inches and a half inside diameter. The paved spaces on each side of the eighteen middle feet are to be coated with broken stones, or well-cleansed strong gravel, up to the footpath or other boundary of the road, so as to make the whole convexity of the road six inches from the centre to the sides of it; and the whole of the materials are to be covered with a binding of an inch and a half in depth of good gravel, free from clay or earth." (See Plate III. Fig. 3.)

The work of setting the paving-stones must be executed with the greatest care, and strictly according to the foregoing directions: when the work is properly executed, no stone can move.

If the work is executed by contract, the inspector should see all the operations as they are going on. He should walk over the pavement when it is completed, and try whether the stones are firmly fixed; and he should not allow any broken stones to be laid on over the pavement till it has undergone an examination of this kind.

In breaking stones, the workmen should be required to break them as nearly cubical as possible. When this rule is not attended to, a great quantity of materials is wasted by first splitting the stones into thin slices, and then breaking them into pieces that are too small, and too thin. If the stones or top metal are not broken very small, the proper degree of smoothness of surface will not be obtained.

When stones are very hard, they never make a very smooth surface. Limestone will make a much smoother surface than whinstone and other harder stones, but they should not for this reason be preferred to harder stones; for these will wear longest, carriages will run lighter over them, and the expense for scraping and repairing will be less. All the soft kinds of stones make heavy roads in wet weather; and in dry weather there will be more friction upon roads made with them, because there will be more dust on their surface.

The breadth of the road which has been described in the foregoing specification of thirty feet, is recommended as fully sufficient for any road, except a road forming the approach to a very populous city. The confining of a road to this breadth contributes very much to preserve the whole surface of it, from side to side, in a good state, and to diminish expense. For when a road is of greater breadth, the scraping and repairing of the excess beyond thirty feet costs annually a considerable sum. Mr. Telford says, on this point, in his second annual report on the Holyhead Road, dated 17th June 1825:—"He" (the surveyor of the Stonebridge and Birmingham Road) "seems to be still too much disposed to prefer a road of a greater breadth than that recommended by me, of thirty feet: he should reflect, that every yard in breadth makes 1,760 superficial yards to be kept in good order in a mile, and therefore that a road of thirty-nine feet wide has 5,280 superficial yards to be coated with materials, and kept clean, more than a road of thirty feet wide. The additional expense of the wider road may be set down at 15*l.* a mile, and this rate for ten miles will make on his road an extra expenditure of 150*l.* a year."

With respect to the convexity of a road, it should be so arranged that it should be slight in the middle. In giving a convexity of six inches to a road of thirty feet in breadth, the convexity at four feet from the centre should be half an inch; at nine feet, two inches; and at fifteen feet, six inches. This will give the form of a flat ellipsis.

The binding, which in the foregoing specification is required to be laid on a new made road, is by no means of use to the road, but, on the contrary, injurious to it. It is, however, unavoidable when a long piece of new road is to be opened; for, without it, the wheels, by sinking into the new materials, would make the draught of the carriages much too heavy for the horses. This binding, by sinking between the stones, diminishes the absolute solidity of the surface of the road, lets in water and frost, and contributes to prevent the complete consolidation of the mass of broken stones.

If the plan here laid down for constructing a road be faithfully executed, it will secure all the objects that can be required. From the moment it is first opened, it becomes daily harder and smoother, and very soon consolidates into as hard a mass as can be obtained by the use of broken stones. The subsoil of the road cannot get into a state of puddle, and rise up and mix with the surface materials, and thus produce those quagmires and deep ruts that are met with every winter on roads made in the usual way.

Although the expense of constructing a road on this plan may seem to be great, on an average of five years, the joint expense of constructing and repairing such a road will be much less than that of constructing and repairing a road made by putting the surface materials on the natural soil, without a paved foundation; for, in point of fact, such a road has usually to be nearly new made every year for some years after it is first opened.*

This method of making roads with a foundation of pavement is described in French works on roads. The following is taken from the *Encyclopédie de l'Ingénieur*, vol. i. page 356:—

"The first course of stones are to be from ten to twelve inches long, and nine inches deep. These are to be set by the hand on the bed of the road,

* The superiority of roads made with a strong foundation has the testimony of those who practically can give the best opinion on the subject, namely, stage coachmen. The following certificate, signed by forty-three coachmen at Bath and Bristol, was laid before the Committee on Turnpike Roads of the Session of 1836:—"We, the undersigned, having been daily or occasionally, during the last nine years, in the habit of driving public stage coaches over the roads leading to and from the cities of Bath and Bristol, do severally declare, as our opinions, that the plan which has been during that time pursued on the Bath turnpike roads, of making and keeping a strong foundation or bottom of large stones, of laying on at once thick coatings of small stones thereon, and of generally preserving a great degree of strength, renders those roads much harder, sounder, and easier of draught than the Bristol roads, where, a very different plan having been followed, the roads are consequently in general weak, unsound, and more distressing to horses drawing carriages; and we therefore think, that any change of plan on the Bath trust would be injurious to the roads and the public."

This certificate was not laid before the House of Commons by the Committee. It was produced by Mr. Wingrove in support of his method of managing the Bath roads.

with their broadest faces down and their pointed ends upwards: the interstices are to be filled with stone chips. The upper course of stones is to be of the hardest kind, broken to the size of an inch cube, on a large stone, as an anvil.

"This course is to be nine inches in thickness, so that the whole thickness of the two courses will be eighteen inches."

The bed of pavement, for the whole width of the road, may, in some instances, be too expensive, in consequence of the difficulty of procuring proper stone.

In such instances it may be expedient to have only the eighteen middle feet of the carriage way with a foundation of pavement. (See Plate III. fig. 4.)

In a district of country where any coarse sort of stone can be got for making a pavement, it will be cheaper to make a road with a pavement and six inches of broken stones than with ten inches of broken stones without a pavement.

The following observations on the expediency of making a paved foundation for a road is taken from a Report of Mr. W. A. Provis, assistant engineer to Mr. Telford, under whose immediate direction all the works on the Holyhead Road, in North Wales, were constructed:—

"The pitching or paving the bottom of a road is a subject which has often been discussed, and, though generally approved of by scientific men, has met with some decided opponents.

"On the old part of the Shrewsbury and Holyhead Road, which extends from Gobowen to Oswestry, as well as in some other places, the foundation of the road had been paved, but in an irregular and promiscuous manner, some of the stones standing near a foot above others, and in some places holes were left without any stones. Upon this a coat of gravel had been laid, and necessarily of very unequal thickness, some of the points of the stones being scarcely covered.

"This road having afterwards been much neglected, the upper gravel, where thin, was worn quite away, or else forced from its bed by being in so thin a coat that it could not bind, and the road's surface was thereby made a continued succession of hard lumps and hollows, with water standing in every hole after a shower, and no means of getting off but by soaking through the road.

"Any stranger, on passing over such a road, would condemn the principle on which it was made. But here seems to be the great error,—that the principle is condemned, instead of the abuse of it. When the paving is put down carefully by hand, of equal or regular height, with no large smooth-faced stones for the upper stratum to slide upon, and the whole pinned so that no stone can move, I have no hesitation in saying that in many cases it is highly beneficial, and in none detrimental. Whenever the natural soil is clay, or retentive of water, the pavement acts as an under drain to carry off any water that may pass through the surface of the road. The component stones of the pavement, having broader bases to stand upon than those that are broken small, are not so liable to be pressed into the earth below, particularly where the soil is soft. The expense of setting this pavement is less than one fourth of that of breaking an equal depth of stones to the size generally used for upper coating; and therefore, in point of economy, it has also a material advantage.

"Mr. Telford in all cases recommends this paving; and the opinion of a man of such experience cannot be treated slightly. He has made more miles of new road than any engineer in the kingdom; and having myself studied for nearly fifteen years in his school, and made a considerable extent of road under his direction, I may venture to say that his practice is not unsupported by experience.

"I should not have said so much on this subject, but from the circumstance of other road improvers having asserted that paving was useless; and I think that assertions on one side should be met with firmness on the other, whenever an important principle is attacked, the correctness of which can be established by reasoning and by facts.

"Whenever any new piece of road has been made, I have taken care that a good bottoming should first be put under the broken stones, because I am satisfied that it makes the road more substantial, and is also less expensive. Some of the new road made on this principle by the commissioners under the act of 55 Geo. III. has now been travelled upon for four years, and its present perfect state, I have no doubt, is owing to the firm foundation which was laid under the broken stones. I must refer to my last report for further particulars of its advantage; but as I did not then notice the comparative expense of the two modes of road-making, it is proper to state it here, in order to justify the course I have adopted.

"Supposing the materials are stone, to be quarried, and carted say a quarter of a mile on an average; that the stoning shall be in both cases sixteen feet wide; that, by Mr. Telford's mode, the bottoming shall be seven inches thick in the middle, and five inches at the sides, and the broken stones six inches in uniform thickness; and that by the other mode there shall be no bottoming, but ten inches in depth of broken stones:

"The expense of a lineal yard on Mr. Telford's principle will be as follows:—

	s.	*d.*
Quarrying $1\frac{3}{4}$ cubic yards of stone (measured on the road), at 1*s.* 8*d.* - -	2	11
Carrying $1\frac{3}{4}$ ditto $\frac{1}{4}$ mile on an average, at 6*d.* - - - - - -	0	$10\frac{1}{2}$
Setting the bottom - - - -	0	2
Breaking the top 6 inches $\frac{8}{9}$ cube yard, at 1*s.* 6*d.* - - - - - - -	1	4
	5	$3\frac{1}{2}$

"The expense of a lineal yard without bottoming would be,—

	s.	d.
Quarrying and carrying $\frac{5}{6}$ of the above quantity of $1\frac{3}{4}$ cubic yards of stone -	3	2
Breaking 10 inches in depth $1\frac{1}{2}$ yards, at 1*s.* 6*d.* - - - - - -	2	3
	5	5

"But if there were plenty of loose stones to be had without quarrying, which is very often the case, the expense per yard, with bottoming, would be,—

	s.	d.
Carriage of $1\frac{3}{4}$ cubic yards - -	0	$10\frac{1}{2}$
Setting the bottom and breaking the top as before - - - -	1	6
	2	$4\frac{1}{2}$

"Without the bottoming it would be,—

	s.	d.
Carriage of $\frac{5}{6}$, the last-mentioned quantity	0	$8\frac{1}{2}$
Breaking 10 inches as before - -	2	3
	2	$11\frac{1}{2}$

"The first of these cases shows a saving of $1\frac{1}{2}$*d.* and the latter of 7*d.* per lineal yard in favour of the bottomed road, a saving which of itself would not weigh much; but, as the bottomed road is the most substantial and durable, it adds one more to its other advantages."

The following extracts from the examinations before the committee of the House of Commons of the session of 1836 furnish a strong corroboration of the accuracy of the foregoing statements, relative both to the utility and the expense of making a paved foundation for a road.

*Extracts from the Examination of Mr. Wingrove, who has acted for upwards of twenty years as the Surveyor of the Turnpike Roads in the neighbourhood of Bath, Taunton, and other populous towns.**

"When you speak of bottoming will you explain what you mean?—I mean taking out the bottom, and laying in a pitching or paving, on the plan laid down by Mr. Telford. I have always thought Mr. Telford's mode of improving roads was the best; in short, his opinions have ever been my guide. I obtained the Holyhead Road reports, and in all my proceedings the principles of road-making and management developed in those reports have been my guide.

"In point of practice, have you been induced to continue to act on that because you have not found other plans succeed?—Yes, from a conviction of the superiority of Mr. Telford's plan, both from observation and experience. I am satisfied that a road bottomed on the plan laid down by Mr. Telford may be repaired at full 15 to 20, or 25 per cent. less than the road where the broken stones are cast or laid down loose, without any prepared bottom. A great deal depends on the subsoil certainly, but even over a rock I prefer laying

* The examinations of Mr. John Provis, Mr. John Macneill, and Dr. Lardner, are given in Appendix, No. V.

down a pavement, if the rock is uneven in its nature. Where you lay it, you make a clear floor.

"Now, in making the bottom you speak of, of what thickness do you make it?—According to circumstances. If the road is one of great traffic, of course I make it rather thicker; if it is one of slight traffic, I make it lighter.

"How many inches should you say?—I should say the bottom should be from five to seven inches.

"Of loose stones or pavement?—I merely take inferior stone; I do not take superior stones for the bottom, but the common brash, or freestone brash; and in a chalk soil, where there is no brash stone to be had, I take pieces of chalk and hand-lay them, so as to secure an equal bottom.

"Is six inches the average width?—From five to seven inches, according to the wear and tear of the road.

"What do you put on the top of that? — That again would depend on the quality of the material.

"Suppose a road of very great traffic?—In a road of great traffic I would put from six to eight inches above for the surface, and in some places, where the surface materials are of very inferior quality, even more.

"And your rough foundation, do you throw that in indiscriminately or pave it?—We hand-lay it; that is, pave it.

"Now, in your experience from having made roads with broken stone upon a paved bottom, have you found the upper stratum of stone crushed in the way that has been supposed?—No; on the contrary, they combine together and form a regular surface; whereas, if on a soft bottom, it is constantly liable to sink in one part and rise in another, particularly if it is a clay soil. There is a new piece of road lately made in my neighbourhood, of nine or ten miles in length: a great part of it was laid down upon clay, and not only upon the solid clay, but even where they have filled up with a softer substratum; and the consequence was, that this winter there was not a heavy carriage that could go over it. I myself rode behind a loaded cart one day, and the cart at almost every step sunk into a hole. The clay rises up and displaces the surface, and they again cover that over with a layer of small stones; after a short time the clay rises again; they lay on fresh stones, and so they go on. There is a piece of road also between Glastonbury and Street, which was laid down fifteen years ago, and when I saw it making, I observed to the man who was at work upon it, that it was a very expensive operation, and if he would look at it in the course of a few years, he would find that he would have to put on as much stone again. I believe there is two feet of solid material upon that road, and yet in the winter it is as soft as a cushion, from the want of a sufficient bottom.

"You have stated you would make your upper stratum of from 6 to 10 inches thick?—Yes; according to the quality of the material and the usual traffic of the road.

"In your experience, with such a stratum, have

you ever known the stones crushed?—No; the road is always more lasting than any other.

"Have you found the road crushed from heavy weights?—Not at all. We have lately completed a small piece of road near Bradford, which was bottomed in that way; it was upon a clay soil, and I threw a little gravel over the clay where it was a solid bed; and on a part where we have raised the road above 15 feet, as it was an object to get it open as soon as possible, I laid a quantity of furze on the loose or artificial soil, and paved upon the furze, putting a little of the old gravel, which came out of the original road, with a very little stone on the surface, and the road now stands perfectly well, though subjected to much heavy wear immediately after it was laid down, and it has now all the appearance of a sound smooth road.

"How much longer would a road of this sort wear as compared with a road made of broken granite without bottoming?—I should say from my own experience, where I have bottomed a road in that way, I have found that one coating has lasted out one and a half upon an adjoining road which had not been so bottomed.

"Is it, upon the whole, your opinion that, where there are plenty of stones, it is cheaper to make a road with six or seven inches of pavement, and six inches of broken stone, than to make a road with ten inches of broken stone without a pavement?—I should say a road with a pavement of six inches at the bottom, and six inches of broken stone above, is considerably cheaper in the end.

"Is it dearer at first?—It is not much dearer at first: if the bottom is five inches thick, we save the breaking of that, and the difference in the breaking of that will be almost, if not quite, equal to the hand-laying of pavement.

"Then it is about the same expense?—Yes; we get pavement laid at a halfpenny the square yard.

"And you use any kind of stone?—Yes.

"Any shaped stone, so that it is not too large?—Yes; the best form we can get.

"Do you ever find these paving-stones rise up to the surface of the road?—Never, if they are properly laid down.

"Are you prepared to state that the expense of laying down this paved road by hand, and the double expense of the materials, is not much more than the ordinary mode of laying down nine inches of broken stone upon the surface of the road as commonly prepared?—There is not a double expense of materials; nine inches of broken stone is not sufficient to form a new road. The materials for the bottom are six inches deep, but that six inches in depth does not require so much breaking as six inches of broken stone would; then upon that I put six inches of surface in ordinary cases, and in extreme cases from nine to ten inches; that makes twelve inches of stone in ordinary cases; six broken and six not broken, instead of nine inches of broken stone; and what we save in breaking we do not lose in materials; and the expense of hand-laying is very trifling."

Extract from the Examination of Mr. Thomas Penson, surveyor of the Montgomery district of roads, who has 350 miles of turnpike roads under his immediate management.

"Will you be kind enough to describe accurately your different modes of forming these roads. As you have 130 miles now in progress, and as it depends on locality, you of course adopt various modes of forming these roads in various parts; will you describe them shortly?—Where the material is of that description that is suitable for pavement, and can be obtained close upon the spot, I lay down a close pavement of stone, with the square sides down, and fill up the interstices upon the upper surface with the chippings of broken stone, so as to form a foundation of seven inches thick in the middle, by four or five inches thick on the sides, for the whole width between the fences. The centre of the roadway, so far as is sufficiently wide for the traffic, is coated with stone, broken so as to press on its largest dimension through a ring of two inches and a half in diameter for seven inches thick, and the space between this metalling and the fences or paths coated with stone of the same description, but not so well prepared. In places where no material of a fit description for a pavement can be obtained conveniently, after having formed a substratum, that is, brought the surface to a proper state and proper convexity, I cover it with broken stones of a uniform size, to the depth of twelve inches; but not in one larger. I do not put the whole body of stone at once in either case.

" From your experience, having practised both, which mode of making a road do you consider the best, the pavement which you spoke of or the broken stone?—I consider, with reference to future repairs, to the solidity of the road, and to its drainage, that a pavement laid as a foundation is an advantageous mode, and ultimately will produce a saving in repairs, where materials can be had suitable to the purpose.

" Will you explain what you mean by a pavement?—A foundation of stone is laid with its square side downwards, which I consider a very material part of the practical arrangement of that system.

" Do you not find a disposition in the large stones which are placed at the bottom to shake up to the top?—It depends in a great measure how they are placed. I have seen a pavement extremely badly laid, and a great deal of pavement I have seen prejudiced by carts going over it for the purposes of carrying materials to the road before any portion of the metalling was put on, by which the formation and advantages of the pavement are completely destroyed; and unless the contractor is very carefully watched he will do that. He will not carry his materials along the lands, or any other route, but he will travel over the road; and, as is invariably the case, he will not metal it as he goes on, but carry his metal over the road upon which he makes this wheel-track, and thereby destroys the formation of the pavement.

" But, without the pavement coming up by the elasticity of the road, do you not see the large stones generally come up?—If the broken material is laid on, and stones of a larger size are laid at the bottom as a portion of the metalling, these large stones will invariably work up.

" Has not the wearing of the smaller materials which coat the road something to do with the larger materials working to the surface?—If the pavement is properly laid, and the interstices perfectly well filled, and a sufficient body of materials laid on that, and of uniform size so as to prevent the lower body from being disturbed, it cannot move.

" What advantage do you find in a pavement over the entire substance being made of broken stone?—I think it is a better drain; and, where the material is suitable to the purpose, it is done at a less expense.

" Then we may comprise the result of your evidence upon the subject of " laid roads," and " broken stone roads," in your reply to this question, where the locality admits of one being cheaper made than the other, then you give the preference to that particular one, but where you have the opportunity of making both you prefer a laid road?

—I do. Over a considerable portion of the road on which I have been employed there are localities where I could not obtain that description of materials which I could advise for a " laid road:" as I before stated, where that could be done, a smaller body of broken stones will be sufficient, and it may happen that the materials for a " laid road" may sometimes be obtained at a greater expense than will answer the purpose.

" Now allow me to ask you, with reference to the cost of these two different species of road; you have stated that you make the roads according to the locality, therefore it is of course according to the materials which are found most convenient; what do you consider the expense of a pitched foundation, with broken stone on the top, per mile, as compared with a broken stone road?—The extra expense of making a road thirty feet wide entirely of broken stone above the expense of making a road of the same width with a " laid foundation," where the materials are of an equally durable quality and are equally convenient to the line, would in my opinion amount to 72*l.* per mile. There are situations where the difference in expense would be greater than this calculation, and there are localities where the materials are of that description, that a road, formed entirely of broken stone, would be less expensive than a road formed upon a " laid foundation."*

* The following are extracts from Mr. Telford's first report on the Holyhead Road, May 1824, and from his sixth report, May 23, 1829: —

These remarks on the necessity of providing a proper foundation for a road will be now brought

---

"Besides the advantages of easy inclinations, ample breadth, perfect drainings, and complete protection, the forming of a smooth hard surface is one of the distinguishing characteristics of this new road. In summer it is not dusty, and in winter it is very seldom dirty; frosts and rains produce but trifling and superficial effects upon it. During the unusually severe frosts of the winter 1822—1823, and the subsequent thaws and heavy rains, the new road was not cut up or rutted in a single instance, though in several parts of the old road, even where it had been put into decent repair, it was too weak to stand such hard tests: it broke up and became as bad as a bog. This breaking-up was not confined to parts of the Holyhead Road, but was the case, and to a much greater extent, on many, and perhaps all the roads of the neighbourhood. In fact it seemed to be almost universally the case on all roads not constructed with strong foundations, and particularly where the substratum was clayey and retentive of water.

"The great superiority which the Holyhead new road evinced at that trying time was doubtless owing to the substantial foundation which had been prepared for it, previous to the upper stratum of broken stones being laid on it. This foundation is a regular close pavement of stones, carefully set by hand, and varying in height from eight to six inches, to suit the curvature of the road: these stones are all set on edge, but with the flat one lowest, so that each shall rest perfectly firm. The interstices are then pinned with small stones; and care is taken that no stone shall be broader than four or five inches, as the upper stratum does not bind upon them so well when they much exceed that breadth. The pavement thus constructed is quite firm and immovable, and forms a complete separation between the top stratum of broken stones and the retentive soil below. Any water which may percolate through the surface is received amongst the stones of the pavement, and runs from them into the next leading or cross drain, and there escapes. Was there no pavement, the water must remain among the to a conclusion by giving a description of the new Highgate Archway Road, made with a foundation composed of Roman cement and gravel.

---

working materials of the road, or rest on the surface of the clay till again evaporated. Should frost succeed, the pavement prevents its acting on the clay, as the water has previously escaped; and as hard stone is not perceptibly altered by frost, it consequently can produce no effect on the surface. Where the water cannot escape, or when, from a want of the intermediate pavement, the frost reaches the clay, which always contains a considerable quantity of moisture, then that moisture or water is expanded by the operation of freezing, and heaves up the whole of the road. A subsequent thaw allows it to subside, but the connection between its parts being disturbed and broken, and the materials loosened, the first heavy weight that passes over them will go quite through, if only of moderate depth, or cut them into ruts when the depth is great. The substratum, too, is reduced to a semi-fluid state, and by the pressure of the hard materials and heavy weights it is forced to the surface as the only means of escaping. If it is attempted to scrape this dirt away directly afterwards, much of the stone or gravel will unavoidably be mixed with it, and also removed. Whilst it remains, the road is scarcely better to travel over than a ploughed field; and when removed, a hole is left on the surface."

Extract from the sixth report.—"In order to ascertain the most effectual mode of rendering the driving-way hard and smooth, I caused an experiment to be made along a quarter of a mile, at the northern extremity of this road (the Highgate Archway Road), by constructing the roadway with a bottoming of Parker's cement and gravel, and with a coat of Hartshill stone laid upon it; and to ascertain what would be the comparative effect of using the same stone on the old surface of the road, I had a large quantity of it laid on between the arch and the Holloway Read. The result is, that between the months of October and March last full four inches of the stone on the old road, between the arch and the Holloway Road, was worn away where eight inches had been laid on, while not one inch was

This road, of little more than a mile and a half in length, was originally made by a private company, at a great expense, owing to the nature of the subsoil, which consisted of sand, clay, and gravel. A tunnel was in the first instance attempted to be made through the hill, which, after having been executed for a considerable length, fell in; the brickwork not being sufficiently strong to withstand the pressure of the clay and sand acting against it. After the failure of the tunnel that

---

worn down where it was laid on the cement bottoming. This result corresponds with other trials where bottoming has been made with rough stone pavement.

"The different parts of the Holyhead Road which have been newly made with a strong bottoming of stone pavement place beyond all question the advantage of this mode of construction; the strength and hardness of the surface admit of carriages being drawn over it with the least possible distress to horses. The surface materials, by being on a dry bed, and not mixed with the subsoil, become perfectly fastened together in a solid mass, and receive no other injury by carriages passing over them than the mere perpendicular pressure of the wheels; whereas, when the materials lie on earth, the earth that necessarily mixes with them is affected by wet and frost; the mass is always more or less loose, and the passing of carriages produces motion among all the pieces of stone, which, causing their rubbing together, wears them on all sides, and hence the more rapid decay of them when thus laid on earth than when laid on a bottoming of rough stone pavement. As the materials wear out less rapidly on such a road, the expense of keeping it in repair is proportionally reduced. The expense of scraping and removing the drift is not only diminished, but with Hartshill stone, Guernsey granite, or other stone equally hard, is nearly altogether avoided."

scheme was abandoned, and open cutting was resorted to.

The road-way was formed by laying on the natural soil a very large quantity of gravel and sand as a foundation, and this was covered, to a considerable depth, with broken flints and larger gravel. The carriage-way, however, notwithstanding this quantity of materials, was so heavy, loose, and difficult to draw over, that many carriages and waggons continued to go over the old steep road, and the company in consequence received little or no return for the capital they had expended in the work. This induced them to try every expedient they could devise to improve the condition of the road, but without success. One of the schemes resorted to was to take up all the road materials, and cover the subsoil with pieces of waste tin, over which gravel, flints, and broken stones were placed. All this, however, was of little or no use: the road continued imperfect, and even dangerous for fast coaches. The Parliamentary Commissioners received complaints and petitions from coach-masters, and other persons who had occasion to work horses over this road, praying that they would cause an examination into the state of the road, and have the defects remedied.

In consequence of these petitions, a Select Committee of the House of Commons, in the year 1817, examined several witnesses upon the state of the road, and, amongst others, a director of the Company, Mr. Hoggart, who stated that the annual expense of keeping the road in repair was 420*l.*

Notwithstanding the investigation which then took place, the road, as may be seen from Mr. Telford's Annual Reports to the Parliamentary Commissioners, still continued in a very bad and imperfect condition.

In 1829, the Parliamentary Commissioners made an arrangement with the Highgate Archway Company, for taking the road under their management, and for borrowing from the Treasury a sum of money to put it into complete repair. In order to accomplish this, several experiments were tried by draining the surface and subsoil, and by laying on a thick coating of broken granite; but from the wet and elastic nature of the subsoil, the hardest stones were rapidly worn away by the wheels of carriages, but much more by the friction of the stones themselves against each other; for, in a very short time, they were found to become as round and as smooth as gravel pebbles, even at the bottom of the whole mass of road materials. It was therefore evident, that to form a perfect road, which might be kept in repair at a moderate expense, it was necessary to establish a dry and solid foundation for the surface broken stones; but as no stones could be obtained for making a foundation of pavement but at a very great expense, a composition of Roman cement and gravel was suggested by Mr. M 'Neill, and this, on trial, was found to answer effectually. The manner of laying down this cemented mass, and constructing the road, is fully detailed in the following evidence, given by Mr. M'Neill before a Select Committee of the House of Commons in May 1830:—

"Are you the resident engineer under Mr. Telford, on the road from London to Shrewsbury?—Yes.

"You conducted the work at the Highgate Archway road?—I did.

"Will you explain to the Committee the expense of the cement composition which was laid on the foundation?—The expense of the cement delivered was 2*s*. a bushel; it was mixed with eight times as much of washed gravel and sand.

"What distance of ground would a bushel so made extend?—Laying on the cement six yards wide and six inches in thickness came to 10*s*. a running yard; but in this case part of the gravel got in taking up the old road was used; if new gravel had been purchased, it would have cost from 12*s*. to 15*s*.: that included the forming the bed of the road, which was done with very great care. There were four drains formed longitudinally, and there were secondary drains running from those to the side channel drains, and those again to drains outside the footpaths, covered with brick, and they all communicated with each other, and discharged the water into proper outlets. On the prepared centre of six yards in width, after it had been properly levelled, the cement was laid on, mixing it first in a box with water, gravel, and sand, in certain proportions: every cask of cement was tried before it was allowed to be used, and when we found it set properly in about fifteen minutes we then used it.

" Did it become hard in fifteen minutes?— Yes, so that we could stand upon it: in about four minutes after being laid, a triangular piece of wood, sheeted with iron, was indented into it, so as to leave a track or channel at every four inches for the broken stones to lie and fasten in.

" For grooves for the stones to bed in? — Yes; and this triangular indent had an inclination of full two inches from the centre to the sides; so that if water came through the broken stones it ran off the cemented mass into the longitudinal drains.

" The cement had that fall from the centre of the road? — Yes, three inches from the centre to each side; that was sufficient to allow the water that percolated through the broken stones to run off.

" What time of year was this composition laid on the road? — 200 yards was done in winter; all the rest in July, August, and September.

" It has been on the road through the last winter? — There has been part of it on since June 1828; nearly two years.

" Have you examined it to see what effect the weather has had on it? — I examined it frequently during the frost, almost every six weeks, and I never came to town without examining it.

" In what state have you found it? — Perfectly hard in every case.

" Not injured by the frost nor the working of

the carriages over it?—Not in the least: there was an injury of six feet square, but it originated from the side banks of the road coming down and bulging it up.

"By under-pressure?—Yes; only six feet square was injured.

"At what rate of expense can a square yard be laid six inches thick of this composition? In the neighbourhood of London it could be laid for about two shillings, according to the locality, per yard six inches thick.

"Was it dearer at Highgate, at the Archway Road?—About twopence dearer than it would in London.

"How many yards wide was it laid?—Six yards wide: the gravel that was found in the road was made use of for part of it; if it had to be purchased, it would have cost *2s.* to *2s. 6d.*

"What state was the road in when it was given into your possession by the Company?—I think it was as bad a road as I ever saw; a coachman could hardly sit on the box when driving along it.

"Was it in consequence of the surface being so uneven?—The surface was uneven in some places; it was not in ruts, but in holes several inches deep.

"What was the surface composed of?—Principally clay, gravel, and sand.

"No body of strong materials?—They could not last on it; they were worn out; an immense quantity had been put on; there were 1,200 cube yards of gravel put on annually. Shortly before we took possession of the road, they put about 800*l*. worth of granite on a small space; the directors said it was necessary to examine the quantity of granite in the road, that we might deduct it from the contractor's prices, as they had laid out out 800*l*. in so short a time; I did in consequence examine it, and opened the road in various places, particularly where they told me it was on, and I could hardly find a stone of it.

"Were not some of the stones worn into a perfect round shape?—Almost every one I found was as round as an egg.

"What, in your opinion, was the cause of these materials wearing out so rapidly?—From the rubbing against each other, and from the weakness of the surface, and the elasticity of the road; in all cases those round stones were at the bottom.

"Was the road damp and wet?—Very wet.

"Had those stones worked themselves down below the furze and tin you say you had to remove?—They were resting on them, and they were quite elastic.

"Can you state to the Committee how many tons of gravel and stone have been laid on the road since it has been put under the care of the Commissioners?—Eight thousand one hundred and forty-six tons of gravel, and 3,614 tons of granite.

"Can you state the number of yards of drains that have been made?—There have been four longitudinal drains made the whole length of the road, besides numerous cross drains, one at every thirty yards; and there have been intermediate small drains every ten yards under the cement; making in all, 12,803 yards.

"What was the reason for making such an unusual quantity of drains?—From the nature of the ground: it was cut through a clay soil, high banks on each side; and all the surface water that came from the slopes and Highgate Hill, came down and rested in the hollows of the subsoil.

"You made use of the cement because you could get no stone to make a pavement?—That was one cause; in that situation it was both better and cheaper than any stone.

"Can you give the Committee any statement of the comparative expense of keeping the road in repair previous to the undertaking of the work by Mr. Telford and since?—Yes, I can; I understand it cost 900*l.* a year; it can now be kept in good repair for much less.

"Have you an opportunity of explaining to the Committee the different degree in which the hard stone wears where laid on the surface of such a road as that was, or laid upon a cement foundation?—Yes, I have; I have tried it repeatedly since the stone has been laid on.

"How many inches did you find of the stone wear out which was first laid on the road before the cement bottoming?—I found four inches gone in one place where it was not on the bottoming, and not half an inch where the bottoming was: the same stone, quartz, was used in each experiment.

" You made the experiment on purpose?—Mr. Telford desired the experiment to be made.

" So that you can state without any risk of error and mistake?—Quite certain; because it was on that the last contract was founded of carrying the work into execution.

" Have you any notion of the saving of horse labour in consequence of the improvement of the road?—I have. I have tried the draught of drawing a waggon over it by the instrument I have invented, and which has been approved of, and ordered by the Commissioners to be used on the Holyhead Road for proving the comparative merit of each part of the road; and I found that it took a power of 56 lbs. on the Archway Road; and I have no doubt whatever that if it had been tried in the first instance, before the improvement was made, that what is now 56 lbs. would have been at least 156 lbs., judging from trials made on roads equally bad; or that fifty-six horses will now do as much work as 156 could before the road was improved.

" Before you were employed by the Parliamentary Commissioners, were you employed in making and repairing roads on the system of putting the broken stone on the soil, or on the surface of the old roads?—Yes, since 1816.

" You have now an opportunity of judging of the comparative wear of materials, whether so made use of, or put on a paved foundation: will you describe to the Committee what, in your opinion, is the general result?—I have attended to it very particularly, and I have no hesitation in saying, that the annual saving from a paved bottom will be one third the expense in any case: either a stone or cement bottom is the same; it is merely the solidity and dryness that is required.

"In your experience and the observations you have made, taking together the first expense of making a road in the ordinary way, by putting the broken stone on the soil, and taking a certain number of years' subsequent expense, say ten years, and, on the other hand, taking the first expense in making a road with a solid foundation, either of stone or cement, with the same sort of broken materials and surface, for the same period of time, which would you say would be the cheapest in the end?—There can be no doubt at all upon that subject: the saving, I think, will be one third; and if you include the labour of horse power that is gained, it will be very considerably more than that.

"In favour of which?—In favour of the paved bottom.

"What would be the saving in the course of ten years in favour of the paved bottom?—The saving in ten years will depend on local circumstances, but in every case a paved bottom will be found the cheapest: in the first place, the breaking of the stone is saved, which, where hard granite is used, is considerable; in the second place, a softer, weaker, and generally a cheaper stone than can be used for the top material may be made use of for the pavement; thirdly, the surface stones are preserved from wear in an extraordinary manner, by resting on a solid firm foundation, instead of mixing with the wet subsoils, and forming a loose elastic road, which takes some years, and wears out many materials, before it becomes hard and solid.

"What do you mean to express by a paved bottom?—Unbroken stone set on edge across the road, six or eight inches deep, and three or four inches thick; set firmly side by side, and the interstices filled up with stone chips packed in with a small hammer. This forms, in point of fact, a regular pavement, but not smooth on the surface, of about seven inches deep, as the road foundation. Five inches at the side, and seven inches in the centre, if the pavement is eighteen feet wide.

"What kind of stone do you use for this purpose?—Sandstone, limestone, or schistus, or such as can be had in the neighbourhood: any stone, almost, will answer that will bear weight and not decompose by the atmosphere.

"Is it not, in many instances, cheaper than any other material you can use?—It is the cheapest in the first instance, because you have not to break the stone; and the subsequent repairs cost less."

Some time after the road was improved as above described, the contractor of the work wrote a circular letter to some of the principal coachmasters and coachmen, requesting to have their opinion of the state of the road.

The following are extracts from some of the answers to those letters, as published in the Appendix to the report of the Select Committee on the Holyhead and Liverpool Road, printed 30th May, 1830: —

"Golden Cross, Charing-cross, 10th May 1830.

"Sir,

"In answer to your letter, dated the 1st instant, respecting the present state of the Highgate Archway road, I have to remark, that it is most surprisingly improved since you have adopted the plan of Mr. Telford; in fact, so much so, that *four horses* can better perform their journey now (both as regards speed and ease,) than *six horses* could do previous to such plan being adopted. And although the weather this year has been much against the roads, notwithstanding that, your much improved plan has well contended against such weather; so much so, that I have not had occasion to require six horses at any time during the last winter, although in various other places I could not possibly proceed unless I employed that number f horses; for instance, from Barnet to Colney, near Ridge Hill, I have constantly gone with six horses, and even then with difficulty.

"B. W. Horne."

"Sir,

"In reply to your letter, I have to state to you the wonderful improvement made on the Archway road. The short time since its completion, and the severe winter it had to encounter, proves beyond doubt the complete success of the plan. Since my commencement in driving on the Birmingham road, it was with difficulty I could get up the Archway Hill, on one side or the other, with *six horses*, but now *four middling horses* are sufficient for any of our loads.

"THOMAS BRAMBLE."

"Lawrence Lane, 15th May.

"Sir,

"In reply to yours of the 4th instant, requesting my opinion respecting the late improvement of the Highgate Archway road, it is with sincere pleasure I am enabled to state that, during the whole course of my experience, I never saw so much improvement in so short a period: in fact, from its being the very worst piece of road between London and Manchester, it is now become, through your exertions, decidedly the best. I was fearful the severe winter we have experienced, setting in so very soon after its being completed, would have broken it up; but I am most happy to say that, during the whole winter, I have not observed a single place where it was the least affected. Previous to this winter, it was all we could do to walk up both sides of the Archway with six horses, and now we can trot up with our heaviest loads with four.

"When I first commenced driving on the above road, we were obliged to keep twelve horses to work a slow coach to Barnet, and now we can work a fast one the same distance with ten horses.

"I think the facts above stated are at once a convincing proof of the full success of the plan you have adopted; and, as one individual concerned, I beg leave to offer you my most sincere congratulations thereon.

"CYRUS J. COATSWORTH."

ROADS MADE WITH A FOUNDATION OF RUBBLE STONES, AND A SURFACE OF BROKEN STONES.

A useful road may be constructed by making a foundation with rubble stones, and laying broken stones or gravel upon them.

The stones should be reduced so as not to have any of them more than four pounds in weight; these should be laid in a regular bed, to the depth of seven inches in the middle and four inches at the sides, supposing the road to be thirty feet in breadth; a coating of small broken stones should then be laid on in the way directed when a pavement is used.

If the subsoil be clay, a course of earth, of any kind that is not clay, of the thickness of six inches, should be laid upon the clay, to prevent it from rising and mixing with the stones.

A road made according to the rules here given will not be a very expensive one: it will answer for cross turnpike roads, and other roads that do not communicate between large towns and collieries.

This plan is much superior to, and not more expensive than, the next plan.

A ROAD MADE WHOLLY OF BROKEN STONE.

A road may be constructed, suitable to light carriages and little traffic, by forming a level bed on the natural soil, and putting upon it a body of broken stones, of twelve inches in thickness in the middle, and six at the sides. The stones should be laid on in successive layers, taking care to let each layer be worked in, and consolidated, before a fresh one be laid on. If the subsoil is clay, a course of earth should be laid upon it, as proposed in the last plan.

Roads of this description are not sufficiently strong for great thoroughfares. This plan, however, having of late been recommended as greatly superior to all other plans by persons who profess to be experienced and scientific road-makers, a number of turnpike trustees have adopted it; but experience has fully established its unfitness for roads of great traffic, in comparison with those made with a proper foundation. The reason is very obvious; for if a coating of small broken stones be laid on the natural soil, the weight of carriage wheels passing over it forces the lower course of the stones into the soil, while the soil is forced up into the interstices between them; the clean body of stones, first laid on to make the road, is thus converted into a mixed body of stones and earth, and, consequently, the surface of the road cannot but be very imperfect as to hardness. It is necessarily heavy in wet weather, on account of the mud the earth makes on its surface; and, in warm weather, on account of a quantity of dry dirt.

A road made on this plan will require, for two or three years after it is said to be finished, the expenditure of large sums in new materials to bring it into any thing like even an imperfectly consolidated state; and, after all that can be done, such a road will always run heavy, and break up after severe frosts; for, as the natural soil on which such a road is laid is always more or less damp and wet, it will necessarily keep the body of materials of which the road is made damp and wet; in consequence of which, the surface of the road will wear down quickly. Hard frosts will penetrate through the materials into the under soil, and, when thaws take place, break up the whole surface.

It is in this way that the ruinous state of most roads, after severe frosts, is to be accounted for.

The following Post Office bulletin, which appeared in the London newspapers of 27th February 1838, will serve as conclusive testimony in favour of strong roads, and against those made "with materials of an uniform size from the bottom," laid on the natural soil; these being the roads to which the Post Office alludes.

"General Post Office, Tuesday, one o'clock.—The mails, during the last two or three days, have again been seriously retarded in their arrival, owing to the heavy fall of snow, especially in the north, and the generally rotten state of the roads. One Glasgow and one Edinburgh mail are still due.

The roads in the west of England and on other lines are so rotten that the wheels frequently sank to the nave."

ROADS MADE WITH GRAVEL.

In a country where no stone can be got for making a road, and nothing better than gravel can be procured, the following plan of employing it may be adopted:—When the bed of the road has been formed, a coating of small gravel should be laid on, four inches thick, over its whole breadth; carriages should then be let to run upon it, and the ruts should be raked in as soon as they appear. When the first coat of gravel has become tolerably firm, another coating, once screened, should be laid on, three inches thick, over the whole surface, and the ruts raked in as before. When this second coat of gravel is consolidated, a third should be laid on, three inches thick: this coat of gravel should be well riddled, and cleansed from all earth or clay, and all pebbles exceeding one inch and a half in diameter should be broken before they are laid on the road. This process should be repeated until there is a body of gravel laid on the road sixteen inches thick in the middle, and ten at the sides, so as to form a convex surface rising six inches from the sides to the centre. The strongest and best part of the gravel should be put on the middle fifteen feet of the road, and the smaller gravel on the sides. In all roads of this description the greatest care must be taken to drain the subsoil by a sufficient number of cross and mitre drains, communicating with the main drains. If this is not attended to, it will be impossible to form a good carriage way.

A road made with gravel in the way here recommended will be much stronger than gravel roads usually are; but it will be much inferior to one made with stone materials. The roundness of the stones prevents them from becoming consolidated by pressure, so as to form a perfectly hard surface; and when the gravel consists of limestone, flint, freestone, sandstone, or other kinds of weak stone, it is so rapidly pulverised, that the friction produced by wheels passing over it adds greatly to the labour of horses.

CHAPTER VI.

FENCES.

In districts of country where stones abound, walls are the best fences: they require less land than hedges, and, when properly built, give a very neat and finished appearance to a road.

The manner of constructing these fence walls will be described in the chapter on road masonry. Where a quick fence is to be raised, the following specification points out every thing that is requisite: —

"A ditch is to be cut and a bank raised, together occupying a space of eight feet in breadth; the ditch is to be on the field side of the bank, to be cut out of the natural ground, four feet wide at top, ten inches wide at bottom, and two feet and a half deep.* The bank is to be four feet wide, and is to be raised by sods, with the green or swarded side out, to the height of fourteen inches above the side channels of the road.

"Two rows of quicks are to be planted on the ditch side of the bank, a bed being first formed for them, of good vegetable mould, fifteen inches deep, and eighteen inches wide. There are to be twelve plants set in every lineal yard: they are to have

* Where the soil is clay the drain should be four feet deep.

good roots, three years transplanted from the quick bed, and of a strong and healthy appearance.

" These quicksets are to be protected by two rows of posts and rails; three rails in each row: the posts to be of good oak, five feet long, five inches deep by three inches wide, with large buts sunk two feet in the ground.

" The rails are not to be more than eight feet long; to be three inches and a half wide by an inch and a half deep, of good elm, oak, or ash timber.

" In each length of rails two centre posts, at least two inches wide by an inch and a half thick, are to be driven into the ground, and fastened to the rails with strong nails.

" Through cuttings, instead of the ditch and mound as before described, a mound is to be raised on each side for the quicks, eighteen inches high, two feet wide at top, and faced with sod on both sides; outlets for the water which collects behind the mound from the slopes are to be formed under it, at intervals of twenty yards.

" The mound to be composed of the best vegetable mould that can be procured.

" The quicks are to be planted in the centre of this mound."

A quick fence may be also raised in the following manner, in dry soils, without any ditch:—

" A border or flat mound, four feet in width, is to be raised on each side of the road: it is to be six inches above the footpath, and twelve inches above the side channels of the road, if there is no footpath. The top of the mound next the fields is to be made with good earth two feet wide, and to the depth of fifteen inches: two rows of quicksets, twelve in each lineal yard, are to be set in the middle of these two feet."

"These quicksets are to be protected by two rows of posts and rails, as before described."

Where timber is scarce, quick fences may be raised in the following manner:—

"A ditch is to be cut five feet wide at top, and eighteen inches at bottom, and four feet deep; the sods where the land is grass to be laid two feet high above the side channels of the road, and the earth taken out of the ditch to be formed into a bank five feet wide, sloped to a breadth of eighteen inches; to be four feet high; two rows of quicksets to be planted on the outward face of the bank towards the field, in the natural soil on the face of the bank."

When a road is formed on a high embankment, a fence may be made according to the following specification:—

"A wall is to be built on each side of the road thirty feet apart, eighteen inches thick at the foundation, and fifteen inches at the top, two feet high above the side channels, and nine inches deep below them; in all, two feet nine inches.

"The stones are to be laid in neat level courses, closely jointed and well bonded on both sides, and to be of a kind that will not decompose by the weather.

"The length of the top stones to be the thickness of the wall, viz. fifteen inches, and from five to six inches deep, to serve as a coping. A mound of earth twenty inches high is to be raised above the wall, with two lines of sods in the front. One row of quicksets, of twelve in each yard, to be planted on the mound. A single rail fence is to made to protect the quicksets: the top of the rail is to be fourteen inches above the mound."

Wherever a road is carried through a deep cutting walls should be built for the road fences, if stones can be procured.

All road fences should be kept as low as possible, in order that they may not intercept the sun and wind, and diminish their effect in producing evaporation.

For this reason, in deep cuttings, the quicks should never be planted at the top of the banks; but always low down, near the side of the road.

All quick hedges along the sides of roads should be clipped every year in the months of August or September. They should be trimmed so as to be perfectly level at the top, and with a regular and even surface on the side next the road.

To ensure regularity in the appearance of the hedges, a line and templet should be made use of in trimming them.

CHAPTER VII.

ROAD MASONRY.

In constructing roads, masonry is used in a great many cases, and too much pains cannot be taken to have it perfect both in plan and execution.

BRIDGES.

In arranging the plan of a bridge for a road, it should be considered how far it may be made subservient to improve the longitudinal inclination of a road, and save perpendicular height.

When valleys are deep and narrow, they may frequently be passed without great inclinations in the roadway, by selecting a proper position, and building high piers and arches for a bridge, if a stream or river is to be crossed, as is usually the case. On the other hand, when the land on each side of a river is flat, the bridge should be kept low, to avoid an inconvenient ascent to the top of it.

The following are the principal objects, with respect to bridges, which road-makers should have in view, viz.: — 1st, the most eligible situation as regards the direction of the road; 2dly, the proper width for the roadway; 3dly, the inclinations of the roadway over the bridge; and 4thly, the number and span of the arches.

The best situation for a bridge, as it respects a road, will evidently be that which preserves the most direct line; but, for the security of the bridge, it is desirable to have a straight reach above it, and no bend near it.

The width of a bridge between the parapets should be regulated by the traffic that is to pass over it. On turnpike roads near large towns the width should be forty feet. On those in the country thirty or thirty-six feet will be sufficient, and on parish roads, twenty or twenty-four feet.

The inclinations of a roadway over a bridge should be very moderate. They should never exceed one in thirty on turnpike roads where it is possible to avoid it, without incurring a great expense in filling for the approaches. The number and span of the arches must depend on various circumstances, which can only be taken into consideration by the engineer on the spot; and even then much more must be left to his experience and judgment than can be derived from any precise rules as to the proper number and size of the arches.

It will be sufficient to state that the main point to be attended to in every case is that the water-way should be of ample dimensions, to allow the whole body of water to pass freely in the highest floods.

For this purpose the bridges below the site of the proposed bridge should be carefully measured, and the effects of floods upon them observed. This will be a good criterion for determining what the water-way should be of the intended bridge.

In making a plan and estimate of the expense of building a bridge, the point to be most fully examined and considered is how to secure a foundation sufficient to support the intended superstructure. An engineer should make accurate borings, to ascertain the nature of the subsoil; and when the slightest defect appears, piling should be adopted.

As the building of large bridges does not come within the scope of this work, directions will be here given for building those smaller bridges only which are required on all roads. The following are specifications of some of the bridges built on the Holyhead Road.

Specification for a Bridge to be built over the main Drain in Maldraeth Marsh.

"The dimensions of the bridge as well as the shape and dispositions of the various parts are described in the annexed drawings.

"The masonry of all the abutments is to be of good hammer-dressed limestone ashler from the quarry of Nant, laid in regular courses, well jointed and bonded, without pinnings in the face.

"The newels at the extremities of the wing walls are also to be built of good hammer-dressed ashler.

"The arch stones and string course are to be of good sound freestone, which may be procured below Maldraeth Bridge and be boated up.

"The parapet coping to be of stones twelve

inches in depth; the stones to be flush with the inner face of the parapet, and to project four inches on the outside. The sharp angles at the top are to be rounded off.

" All the rest of the masonry to be of good sound rubble masonry, built with stones from the Nant quarry, or others equally good.

" The whole of the masonry, except the inverted arches and water wings, is to be set in mortar composed of one part of good well-burned lime, and two parts of clean sharp sand, to be well mixed and incorporated together.

" The inverted arches are to be of sound limestone, set on edge without mortar.

" The timber for the platform to be either of oak, elm, beach, or fir.

" The approaches are to be embanked at one in thirty from the ordinary level of the road; they shall have a turf mound on each side; the breadth to be thirty-two feet in the clear. The embankments to be carried up in courses not exceeding three feet each in height, and the side slopes to be two to one.

" The metalling over the roadway to be as described in the general specification.

" There are to be water wings built both above and below the large arch of the bridge of dry limestone, each twenty-five feet in length, built vertical at the abutments, and gradually increasing in batter till they are one to one at the extremities. They are to be founded six feet below the springing of the arch, and to be built up to the level of the springing. They are to be three feet thick at bottom and two feet at top.

" The inverted arches are to extend twenty feet above and below the bridge."

Specification for the Bridge at Pont-y-padoc. Fifty feet span.

" A new bridge is to be built over the river agreeably to the drawings which are hereunto annexed. (Plate IV. fig. 1.) The masonry of the abutments is to be of good hammer-dressed stones laid in regular courses, well bedded and jointed, and without pinnings in the face. The wing walls, spandrells, and parapets, all (but the coping) are to be of good rubble masonry, composed of stones that are sound and will stand the weather, also without pinnings. The arch stones, the string course above the arch, and the parapet coping, are all to be of stone from the quarry above Rhydllanfair, or of equally good quality.

" The stones for the parapet coping are to be laid flat, well jointed, not less than three feet in length and nine inches in thickness. All the masonry above described to be laid in good lime and stone mortar.

" There are to be water wings above and below the bridge, to connect the abutments with the solid rock; they are to be built as high as the springing of the arch, and the total length of the four walls is not to exceed eighty feet. These walls to be built of good lime mortar.

"The filling in between the wing walls, spandrells, and for the approaches, to be brought up in regular layers, or courses not exceeding five feet in height each.

"There must be dry breast walls on each side of the road at each end of the bridge, and parapets in lime-mortar on the top of each for the whole length of the 139 yards, excepting the space the bridge occupies. The line of the top of the parapet at the N. W. end of the bridge shall be an uniform incline from the point whence the lot commences to the wing walls of the bridge. The whole length to be metalled as per general specification."

Specification of a Bridge built over the Ellesmere Canal, on the Holyhead Road.

"The dimensions of the bridge are described in the annexed drawings.

"This bridge is to be built of good sound freestone or limestone; but the two sorts are not to be mixed, and all to be set in mortar composed of one third of well slaked lime, and two thirds of clean sharp sand, well wrought together.

"The abutments, wing walls, spandrells, water wings, and parapets, all (but the coping) are to be of good coursed rubble work; the beds and joints fairly and squarely wrought with the hammer or pick, so as not to require pinnings on the face.

"The copings of the water wings and parapets, arch, string course, and caps of newels, to be neatly drafted round the bed and end joints, and the rest of the face to be neatly dressed with a pick point. The bed and end joints to be truly wrought.

" The coping of the wing walls to be continued through the abutment, so as to prevent the boats striking the face of the wall.

" The wing walls are to have a curved batter on the face, of one and a half inches to the foot. They are to be vertical behind, with two six-inch onsets, the lowest two feet above the springing of the arch, and the upper one six inches above the crown of the arch, so that the inner line of the foundation will be one foot nearer the centre of the bridge than the inside of the parapet.

" The water wing walls are each to be forty feet long from the face of the arch; to be two feet thick at top, and three feet at bottom; each to splay back so as to cut into the land at top bank level, and to have a gradually increasing batter on their faces, so as to be not less than four inches to a foot at their extremities."

Specification for the Bridge built on the Coventry New Road, Plate IV. fig. 2.

" The ground where the bridge is to be erected is to be excavated ten feet deep, for the purpose of laying a foundation of timber for the abutments and wing walls.

" In the space thus opened, 102 piles are to be driven; they are to be six feet long: sixty-six of them, which are to be under the abutments, are to be nine and six inches; the remaining thirty-six, which are under the wing walls, are to be six inches square; they are to have a shoeing of three eighths of an inch iron, nine inches long, and two inches broad. The upper end of the piles are to be cut into a tenon to be inserted in the sills.

"The sills are to be twelve inches by nine inches, to be laid perfectly level, and their upper surface eight feet four inches below the ground line: the spaces round the pile heads and between the sills are to be firmly packed with rubble stone and grouted. A four-inch sheeting of good elm or battens is to be laid over the sills, to be thirty-three feet long, seven and a half feet broad for the abutments, and five and a half by eight and a half for each wing wall; it is to be laid close, evenly bedded, and spiked to the sills. The upper surface is to be eight feet below the ground line. On this platform the abutments are to be built of stone masonry in level beds and grouted; they are to be seven feet thick at the foundation, and diminishing by two offsets of six inches each to six feet at the springing, which is to be at eight feet above the top of the platform: they are to be faced nine inches thick with blue metal bricks laid header and stretcher in clean joints. The wing walls are also to be of stone masonry, to be six feet thick at the platform next the abutments, and diminishing to five feet three inches at eight feet from the abutments; this thickness is to be continued to the height of five feet three inches, where their length is increased eight feet nine inches; at this height there is to be an offset of one foot, and the work brought up from this to the ground line four feet three inches thick.

"The arch is to be thirty feet span, rising ten feet; to be built of the best blue brick that can be procured; it is to be one foot ten inches and a half thick at the springing, and to continue of the same thickness from nine feet from the springing, where there is to be an offset of four inches and a half; from thence to the crown it is to be eighteen inches thick; the headers of the arch to show only eighteen inches all round. The arch is to be turned on a centre, constructed of good timber, to be approved of by the engineer.

"The backing of the arch is to be of good stone masonry three feet six inches at the springing; at the height of two feet above the top of the abutment, there is to be an offset of one foot; at the height of four feet, there is another of a foot; and at the height of five feet to slope to the back of the arch. The water wing walls are to be ten feet long; they are to be founded at the depth of eight feet below the ground line, and to be built of stone masonry two feet thick at the bottom, and diminishing by an offset of six inches at the height of three feet, and to continue of that thickness to the top; they are to rise to the ground line, and to be coped with sod; the splay to be three feet in each, or to be thirty-six feet in the clear at the extremities.

"The spandrell and wing walls above the ground line are to be of blue brick masonry, to be three feet thick, with nine-inch counterforts, founded on the stone masonry before described; at the height of five feet, there is to be an offset of three inches, and the remaining thickness of two feet nine inches to be continued to the road line at this height. There is to be a string course of white freestone, six inches thick, one foot on the bed, and projecting two inches over the face of the work; the under edge to be one foot six inches above the upper line of the arch at the crown: a parapet wall is to be built on each wing wall, and over the arch, of fourteen inches blue brick work, three feet six inches in height, and coped with stone three feet in length, six inches thick in the centre, and four inches and a half at the sides, which are to be flush with the brick work.

" The stones are to be well bedded and jointed with four-inched cast-iron dowels, and the joints run with cement. The whole of the works and materials are to be approved of by the engineer. The mortar to be made use of is to be composed of three parts of sand to one of the best Newbold lime, to be properly worked and used while fresh. All the face work to be laid in mortar composed of two parts of Newbold lime to one of sharp clean sand."

RETAINING WALLS.

Where the natural surface of a country is very rugged and precipitous, it will frequently be necessary to build retaining walls.

The following is a specification for building a

retaining wall on part of the Holyhead Road, in North Wales (Plate IV. fig. 3.): —

"On sloping ground there must be a retaining wall along the upper side of the road eighteen inches wide at top; its foundation to be laid at least four inches below the bottom of the side drains, and is to be carried up, so as to intersect the slope of the bank, when falling at the rate of two horizontal to one perpendicular; and the slope is to be formed in this manner for at least one yard from the back of the wall by means of swarded turf or stone pavement. The face is to have a curved batter at the rate of one inch and a half in every foot from the top: the back may be in the shape of a rough dyke wall; but every one of the back stones are to be regularly connected with the body of the wall, and not to depend upon the earth behind them."

If a retaining wall be built of brick, the thickness at top should be one brick, or nine inches, and it should increase in breadth by onsets of half a brick at every eight courses to the level of the road, below which the thickness for the stepping of the foundation should increase half a brick at every four courses to the bottom. All the walls of this description should batter in a curve line on the face at the rate of one inch in every foot.

BREAST WALLS.

These walls are necessary for supporting earth or other materials used for forming a road; they should be built in the same way as retaining walls, and should increase, from one foot six inches in breadth at top, at the same rate as has been described for retaining walls.

These walls should have a strong coping of large stones, set on edge in mortar of the best description.

The following is a specification of a breast wall built across a very deep hollow in North Wales, on the Holyhead Road (Plate IV. fig. 4.): —

" Across the hollow there is to be a breast wall built, in good lime and sand mortar, along the foot of the lower slope of the present road, or thirty feet distant from the retaining wall. This breast wall is to be two feet and a half thick at top, and to increase in thickness downwards at the rate of two inches and a half for every foot of depth, by a regular batter on the outside. There is to be a four-feet parapet wall on the top, two feet thick at the bottom, and eighteen inches at the top."

FENCE WALLS.

These walls may be be built without mortar, if the stones are flat bedded.

As their stability depends upon the workmanship, great care should be taken to have the stones properly selected, and laid in a correct and regular manner.

A coping should be made on the top of these walls; it should be of large stones set on edge, and laid in good mortar.

When walls are used for fences on embankments,

they should always be built with mortar, or otherwise the shaking of the road will in most cases loosen and throw them down.

The following is a specification for stone fences on the Holyhead Road:—

"The stone dykes are to be four feet six inches high above the side drain; they are to be, when placed on breast walls, two feet thick at the bottom, and sixteen inches at the top; and where there is no breast wall below them, they are to be two feet six inches thick at the foundation, and one foot six inches at the top."

CROSS DRAINS.

Cross drains should be built of good masonry eighteen inches in the clear. (Plate IV. fig. 5.)

They should be continued under the fences into the ditches on each side of the road. When made of stone masonry the side walls should be sixteen inches thick, faced on both sides, eighteen inches high at the upper end, and twenty-three inches at the lower end. The top of the walls to be level, and the bottom of the drain to have an inclination of one inch in every ten feet. The stones at top on which the covers are to be laid are to project about two inches and a half into the open space on each side, leaving about thirteen inches clear between them: the covers to be stone not less than four inches thick and twenty-seven inches long; they should be neatly jointed and closely laid together, and properly bonded on the side walls: a concave pavement of stones, not less than five inches deep, should be laid between the side walls. The body of the building should be placed so low as to admit of six inches of earth to be laid between the cover of the drain and the bottom course of the road materials, without elevating the surface of the road.

The ends of the cross drain must be secured with a strong pavement, four feet three inches by two feet three inches; the paving-stones below the discharging end should be of large stones, sunk so deep as to secure the whole from being injured by the current of water.

When a cross drain is connected with a water-course, the upper end should be secured with wing walls, at least five feet in length, and there should be similar walls at the lower end. These wing walls should be covered with two rows of swarded turf, the lower one with the swarded side down, and the upper one with the swarded side up.

The following is a specification of a cross drain, five feet diameter, built on the Holyhead Road:—

"The arch to be hammer-dressed coursed work, and the rest of good sound rubble-work. It is to be in length the full breadth of the road and dykes. The faces to range with the faces of the breast walls, and the dykes to be continued over them.

"Water wings are to be built into and extended from each abutment for eight feet in length, and to splay back to eight feet apart at their extremities. They are to be founded at the same depth as the abutments, and be carried up to the level of natural ground. A stone pitching to be set between the abutments and water wings; to be set endwise to the streams, and be firmly secured at each extremity. Except the stone pitching, the whole is to be built in good lime and sand mortar. The thickness of the water wing walls to be the same as specified for the breast walls."

Specification for a Three-feet Stone Drain.

"The arch to be hammer-dressed, and the rest of the masonry good sound rubble-work. The abutments must be continued as water wings above and below the arch, for five feet in length, and be splayed back at their extremities; to be founded as low as the abutments, and rise to the springing of the arch. A dry stone pitching to extend under the arch and between the water wings.

"Except the pitching, the whole to be set in good lime and sand mortar up to the level of the roadway; to be the full length of the breadth of the road and dykes. The faces to range with the faces of the breast walls, and the stone dykes to be continued across the arch in the usual manner."

INLETS.

The water from the side channels of a road should be introduced into the cross drains by side openings or inlets; these should be built with stone masonry, and be ten inches by sixteen inches, and covered with sound flags, at least twenty-six inches long and sixteen inches broad, and two inches and a half in thickness. The top of these covers should be six inches above the level of the sides of the channels, and the whole of the inlet should be built outside of the side channels, as shown in Plate IV. fig. 6.

Inlets may be made along the side channels, and covered with iron grates eighteen inches or two feet square: the bars of the grates should be three quarters of an inch broad, two inches deep, and one inch apart, if made of cast iron: if the grates are made of wrought iron, it is usual to set the bars in an oak curb; but the cast-metal grates are found to answer better, particularly if bedded in stone or on brick curbing. In some situations it is found necessary to leave an opening or inlet under the footpath, as first described, as well as the grate, to allow the water to get off in thunder-storms: a provision is also sometimes made in the casting, to allow the grates to be turned up on a hinge, in case of sudden and large runs of water.

OUTLETS.

Outlets are necessary to receive and carry off the water from the side channels of the road. These outlets may be built of brick or stone: in most cases they should be about one foot square; when they are for the purpose of carrying the water from the channels into the side drains, on grounds nearly level, they may be made of large six-inch diameter tiles or iron pipes. There should be an outlet at the end of every cutting, to allow the water that collects in the side drains through the cutting to run off to the side drains before it reaches the embankment. For want of this precaution, embankments frequently suffer very much.

DEPÔTS.

Depôts should be made on the sides of all roads for holding materials.

The best form is that which will serve to measure the quantity of materials, as well as to hold them.

The back walls should be twelve yards long, and the two side walls each two yards and a half at the bottom, and to slope at an angle of 45° to the top: the height of the back and sides should be three feet. A depôt of this form and dimensions will hold twenty-four cubic yards of materials; and four of these depôts on a mile, at 428 yards apart, will contain 100 cubic yards.

The depôts should not in any case be placed farther from each other than a quarter of a mile, so as to admit of moving the materials in barrows. This method is better calculated for constant repairs, than that of drawing them with carts and horses.

Specification for building Depôts of Stone Masonry.

"Four depôts are to be built, on each mile of road, in such places as may be pointed out by the engineer or his assistant: they are to be built with stone and lime; twelve yards long in the clear, three feet high above the side channels of the road, and to be founded as low as necessary below that, to give stability to the work; the ends to be two yards and a half in the clear at the bottom, and to rise to one yard and a half at top; the thickness of the work to be eighteen inches throughout for the height of three feet; the work under that to be two feet thick. The top of the back, sides, and slopes to be coped with large stones, set on edge, and laid in good mortar.

"The bottom is to be flagged with sandstone, in the rough, neatly jointed and evenly bedded.

"The back and ends of the depôts to have a mound of earth thrown up against them, eighteen inches high on the outside, and eighteen inches or two feet on the base, rounded off on the top, and faced with sod if necessary, and the regular quantity of quicksets planted in it, which are to be protected by the field row of posts and rails before described: a tile drain to be laid in front of the depôt; it is to be thirteen yards long and ten inches in the square." (See Plate IV. fig. 7.)

Specification for building Depôts with Bricks.

"Four depôts for holding repairing materials are to be erected, in each mile of road, in such places as may be pointed out by the inspector: they are to be built of brick, twelve yards long in the clear, three feet high above the side channels of the road; the foundations to be deep enough to give stability to the work; the ends to be two yards and a half clear at the bottom, and to rise one yard and a half at top; the thickness of the back and ends to be nine inches. The upper course of bricks are to be laid on edge in cement; the ends are to be secured by an oak post driven three feet in the ground, and rising one foot above the surface; the bottom is to be paved with brick on edge, and an oak plank is to be set in front the whole length of the depôt, and flush with the upper surface of the pavement; it is to be three inches thick and six inches deep, secured at the ends by the oak posts before mentioned, and strengthened in the middle by a post driven two feet into the ground, and of a scantling not less than four inches by six inches.

"The back and ends of the depôts are to have a mound of earth raised behind them eighteen inches deep and two feet wide, and faced with green sod; in this mound the quicks are to be set, which are to be protected with one row of posts and rails, as above described, set on the field side. A drain is to be made in front, of brick or tiles, so as not to interrupt the passage of the water in the inner side drain."

TOLL-HOUSES.

Toll-houses should be built in a strong and substantial manner, and made suitable and comfortable for the persons who are to inhabit them. Many instances might be mentioned in which the tolls on a road have been much increased by building good houses. The following are the specifications for building toll-houses on the Holyhead Road.

Specification for building a Toll-house at Llanfair, in the Island of Anglesea. (Plate IV. fig. 8.)

"The toll-house is to be built at the precise spot now marked out on the ground, and to be in shape and dimensions agreeable to the above drawings.

"The masonry to be of good sound rubble-work, except the plinth, steps, and sills, which are to be of good hammer-dressed freestone, or slate. The whole to be set in good lime and sand mortar.

"The sills of the door and window frames to be of oak; the rest of the frames, and outer wood-work, to be of Baltic fir, except the posts of the portico, which are to be of sound round oak. All inside timber-work, to be of Baltic fir.

"The scantlings of timber and description of workmanship to be similar to those of the toll-house at Llandegai.

"The roof to be covered with slates, and the hips, ridges, and gutters to be covered with lead eighteen inches wide, and not less than seven pounds to the square foot.

"The inside wall and ceilings to be plastered three coats, and set. The under side of the portico, and the projection of roof, to be also ceiled and plastered, and faced with a three and a half inch fascia board. The outside to be roughcast and coloured.

"The portico to be paved with pebble, with a hammer-dressed plinth, for the posts to stand upon, at least twelve inches wide. The octagonal lower room and the wash-house to be paved with tiles.

" All woodwork to be painted three times in oil; the inside works finished white, with doors and skirting oak colour, and the outside work dark green.

" There are to be proper grates, with slate chimney-pieces, to all the rooms.

" A garden is to be fenced round on three sides, each of twenty yards, with a walk of the same description as those on each side of the new road.

" In this garden a privy is to be built, with proper roof, dome, seat, &c. complete. There are to be two wrought-iron toll-gates hung, one across the road to Plas-newydd, and the other across that to Holyhead. There are also to be two turnstiles; the posts and rails to be of sound oak; to be painted three times in oil, white. The contractor to find all the materials and the labour."

Specification for building a Toll-house at Shelton, in the Parish of St. Chad, in the County of Salop.

The toll-house to be built on the spot marked out by the engineer, and agreeably to the drawings.

" The whole of the walls, except the plinth, steps, and window sills, are to be of good sound brick-work, to be laid solid in good mortar, composed of lime and sand; and the outer joints to be neatly struck with the trowel. The plinth, steps, and window sills to be of neatly tooled freestone.

"The rooms numbered 1. and 2. on the plan are to be floored with paving tile, well bedded in mortar. Those numbered 3. and 4. are to have joists of batten fir, sixteen inches from centre to centre, seven inches deep, and two inches and a half thick, and to be covered with wrought-inch board of fir or poplar. All the ceilings to have joists not less than three inches by two inches, and the roof to have rafters of the same dimensions, with purloins four inches square under the whole of them. The ceiling, joists, and rafters to be sixteen inches from centre to centre. Hips and ridges to be one inch and a half by eight inches, and raised to receive the lead. The wall plates to be four inches and a half by two inches and a half.

"The roof to be covered with good slate, nailed on two-inch sawn lath, and to be rough-rendered inside. All the hips and ridges to be covered with lead, fifteen inches wide, and not less than five pounds to the foot, and valleys not less than six pounds to the foot.

"All ceilings to be plastered on heart lath, two coats, floated and set.

"Walls to have two coats, rendered and set. The under side of the portico and projection of the roof to be also ceiled and plastered, and faced with a three inch and a half fascia board.

"All ceilings and plastered walls to be white-washed, to have six-inch skirting round all the rooms, of fir, one inch thick.

"Outer door frames to be of oak, four inches square. The front door to be two inches thick, six-panelled, square, and rusticated. The back door to be of one inch and a half deal, ledged, ploughed, tongued, and beaded.

" Inner door frames to be of fir, not less than four inches by three inches. Inner doors of one and a half inch fir, six-panelled. Window frames to have oak sills, and the rest of the framing to be of fir, glazed with best second diamond fashion in lead, with quartered oak casements, and iron saddle bars.

" There is to be a five eighth of an inch beaded angle staff at every external angle of the plastering.

" All timber work, except where otherwise expressed, to be of Baltic fir.

" The porch to be supported with oak or larch poles. All the timber work usually painted to be painted three times in oil. The inner doors and skirtings to be finished chocolate colour; the rest of the interior to be white; all the outside to be painted dark green.

" Grates must be fixed in the fireplaces, each of which must have stone hearths, jambs, and mantels.

" All doors and windows to have proper locks, hangings, and fastenings; and the whole house to be finished in a complete and workmanlike manner.

" A privy must be built near the house, with proper seats, roof, door, &c. complete, to be placed in such situations as may be pointed out by the commissioners' engineer or his assistant."

Specification of a Toll-house, built near Coventry, on the New Road. (Plate V. fig. 1.)

" The precise situation to be marked out by the chief engineer of the parliamentary commissioners, or the resident engineer: it is to be constructed agreeably to the annexed plan and following particulars:—

" The excavation of ground for the walls to be of sufficient depth to obtain a solid foundation; the whole of the walls to be formed at the same depth, and the earth to be firmly pinned up to them after they are built."

Brickwork.

" The exterior walls are to be two bricks in thickness at the foundation, and be continued at that thickness up to six inches above the outside ground line, where there must be an offset, on the inside, of two inches and a quarter: at six inches above the floor line, or twelve above the outside ground line, there must be another offset on the outside, so as to bring all the walls to be a brick and a half in thickness, which thickness is to be continued to the roof. At the height of nine feet above the floor line, and ten above the outside ground line, all the brickwork to be of sound stocks laid in well-wrought mortar, composed of one part of good lime and two of clean sharp sand; the two middle rooms to be laid with good paving bricks; the chimney-shaft above the roof to be half a brick in thickness, finished with proper coping; all the flues to be pargetted smooth."

Carpenters' Work.

"For the roof, which is to rise a quarter of the span, all the rafters are to be four inches deep, two inches in thickness, and laid at the distance of sixteen inches from centre to centre; the hip, ridge, and gutter pieces to be six inches deep, and one inch and a half thick; the ceiling joists to be the same, laid at the same distance, and framed to the rafters.

"The two bedrooms to have flooring joists of the same dimensions and distance as the rafters, and laid with inch deals, folding-wrought, and nailed. The front and back door sills are to be of oak, nine inches by six inches, and three inches in the walls. Window sills to be of oak, nine inches by three inches, with a weathering of one inch.

"The front and back door frames to be of oak, four inches square, framed, rebated, and beaded.

"Interior door frames to be of fir, four inches by three inches, framed, rebated, and beaded. All window frames to be of fir, three inches square, with two mullions in each, three by two inches, rebated and moulded, for lead lights; all the rooms to have six-inch ovolo skirting, one inch thick, with proper grounds; all angles inside to have a three-quarter inch bead, with quirks wrought in plastering."

Slaters' Work.

"The roof to have sawn laths, two inches by half an inch, and to be covered with the best countess slates, laid in a proper manner, and fixed with clout nails. A two-inch beaded fascia to be run along the ends of the rafters, which is to be bracketed underneath. The projection to be fifteen inches; the slating to be pointed in the inside with plaster mortar; the wall top to be beam-filled."

Joiners' Work.

"The front door to be six-panelled, bead, flush, and square, two inches thick; the back door to be four-panelled, square, and rusticated, one inch and a half thick. The interior doors four-panelled, square, one inch and a half thick; all the wood-work (except where specified to be of oak) to be of sound Baltic fir. All framed and outside work to be of yellow fir, as well as flooring boards."

Plasterers' Work.

"All the walls inside to be rendered, floated, and set, also the outside of the blank windows. All ceilings to have two coats on sound heart laths, and be properly set. All the plastering to be afterwards whitewashed two coats."

Plumbers', Painters', and Glaziers' Work.

"All hips, ridges, and gutters of the roof to be covered with lead, fifteen inches in breadth, and not less than 7 lbs. to the superficial foot. All door frames, window frames, skirtings, angle beads, fascia, and all wood and iron work whatever, to be painted three coats in oil; the doors and skirtings to be finished an oak colour. The windows to have small diamond glass, in lead and iron casements."

Ironmongery.

"There are to be iron frames for lead lights for all the windows; the middle to be an opening casement, with proper hinges and fastenings. The front and back doors to be hung with four-inch best buts, and each to have a strong rim lock, and two bright bolts; all the inner doors to be hung with three inches and a half buts, and each to have a strong rim lock.

"Plain square grates to be fixed in each of the fireplaces."

Miscellaneous Matters.

"All the fireplaces to have neat, plain, square stone jambs, lintels, and mantels.

"A toll board to be made and painted with the rates of tolls, and fixed up where directed by the engineer.

"The contractor to find all materials and labour, and finish the whole to the satisfaction of the before-mentioned engineer, on or before the day of ."

TOLL-GATES AND BARS.

A toll-gate should never be placed on a hill or at the bottom of one. When going up hill, the horses must make a great exertion to put a carriage in motion after being stopped at a toll-gate. Many fatal accidents have occurred from having toll-gates just at the bottom of hills.

When circumstances render it unavoidable that a toll-gate should be placed at the bottom of a hill, the gateway should be very wide. If a single gate be used, it should not be less than fifteen feet in the clear: but, in such a situation, it is much better to make double gates, meeting in the middle, without a centre post; by these means an opening may easily be had of from twenty-four to thirty feet in the clear.

Toll-gates should be painted white, to make them more easily seen in the night-time. They are frequently made too high. When this is the case, they are expensive and unsightly, and their additional weight acts as a powerful leverage in straining and pulling the hanging post out of its place.

The toll-gates erected by the parliamentary commissioners at South Mims, and on the Coventry road, are only four feet six inches high: they open to sixteen feet in the clear (Plate V. fig. 2.); they are hung on posts made of the best oak, sunk five feet in the ground, and secured by brickwork and struts; there are also two bars passing diagonally from post to post, by which means they are firmly braced together.

These gates are hung on Collinge's patent hinges, which are particularly fit for this purpose; they run about five feet along the upper and under rail of the gate, and are connected by a diagonal piece of metal, carried from the bed of the lower hinge to the point of the upper one, in order to prevent the gate from sinking. The balls of the hinges are cast with the caps and plinths of the posts, so that the posts are not weakened by holes or mortices, as in the usual manner of hanging gates. The caps and plinths of metal are also a great security to the posts, by preserving them from the effects of the weather, and by preventing the wheels of carriages from chafing their angles.

Flapping posts are set in the ground at proper places, to prevent the gates from opening too far, and straining the hinges; these posts are about two feet and a half above the ground, and two feet in it. Catches or clicks are let into these posts, to hold the gates open when thrown back; these catches project about two inches from the side of the posts, and turn on a pin within the post, the inner end of the catch being made heavier than the outer, and always throws that end up, and by that means it takes hold of the bottom of the lower bar of the gate, by a notch cut in it for that purpose: by making the catches in this way, they are out of the reach of injury. In the common way they are put on the top of the posts, from which they project six or seven inches; in consequence of which they are frequently torn off by wheels of carriages and waggons.

LAMPS.

All toll-gates should be well lit; and for this purpose nothing is better than a lamp made similar to the best coach lamp, with powerful reflectors, and large air holes. These lamps are found to be economical, and to answer every purpose; they should be about ten inches high, and seven inches square in the clear; they cost about 1*l*. 7*s*. each.

MILESTONES.

Milestones are convenient and agreeable to travellers, and useful in enabling coachmen to keep their time with accuracy. They are also serviceable in assisting road surveyors in laying out and measuring work. They should be made of very hard stone of a light colour, and should be much larger than they usually are, in order that they may be readily seen, and have space enough for large figures; for if they are not large it is difficult when going fast to read them. A drawing of a proper shaped milestone is given in Plate V. Fig. 4.

The figures XIV. in this drawing show the distance from London, but though large, they are not large enough, and therefore it would be better on all main roads from London to have on the plate only the distance from London, in the largest figures the plate will admit of.

CHAPTER VIII.

MANAGEMENT OF ROAD WORKS.

When a new road is to be made, as soon as the precise line of it is finally determined upon, the following circumstances should be particularly attended to.

I. Drawings should be made to show, 1st, the natural surface of the ground; 2ndly, the longitudinal inclinations of the proposed road; 3dly, the slopes of the cuttings and embankments; 4thly, the form of the bed of the road and footpath; and, 5thly, the courses of materials to be laid on, and the thickness of each course.

Drawings should also be made, describing the plans of the bridges, culverts, cross drains, inlets, outlets, depôts, and fences which are required to be made.

II. A specification should be prepared, to explain in detail the precise method of executing every part of the work.

III. After the specification has been settled, an estimate should be made of the expense to be incurred.

The following is the estimate for making part of the Holyhead Road, near Coventry: —

FIRST DIVISION: *Seven Stars to Coventry.*	£	s.	d.	£	s.	d.
To removing 26,790 cubic yards of earth, at 9*d*. - - -	1,004	12	6			
To 1,882 lineal yards road-making, at 15*s*. 6*d*.* - - -	1,458	11	0			
To 1,882 lineal yards of fencing, at 5*s*. - - -	470	2	0			
To drains - - -	50	0	0			
To forming - - -	100	0	0			
To sodding - - -	100	0	0			
To depôts - - - -	48	0	0			
To bridge - - - -	670	0	0			
To side roads - - -	20	0	0			
To gates - - - - -	8	0	0			
To 5,200 cubic yards extra embankment, at 6*d*. - -	130	0	0			
To extra carriage of Hartshill stone	81	0	0			
				4,140	5	6
SECOND DIVISION: *Coventry to Allesley.*						
To removing 38,990 yards of earth, at 9*d*. - - - -	1,462	2	6			
To 3,760 lineal yards road-making, at 15*s*. - - - -	2,820	0	0			
To 3,760 lineal yards of fencing, at 5*s*. - - - -	940	0	0			
To drains - - - -	95	0	0			
To forming - - -	100	0	0			
To sodding - - -	156	0	0			
To gates - - - -	16	5	0			
To side roads - - -	19	0	0			
To Allesley bridge - -	82	3	6			
To skew culvert - - -	150	0	0			
To depôts - - - -	96	0	0			
Miles. Yards.				5,936	11	0
Length of First Division - - } 1 122				10,076	16	6
Length of Second Division - - } 2 240						
Total Length - 3 362						

* The stones used for this road were brought eight miles, from the quarries at Hartshill near Nuneaton.

IV. The next step to be taken is to make a contract for executing the work.

Contract work is commonly supposed to be preferable to other work, for no other reason than because it is the cheapest, but experience shows that, when it is properly regulated, it is by far the best mode of securing sound and durable work. This, however, will not be the case if the contracts and specifications are prepared by unskilful and inexperienced persons, if inspection is omitted, and if the contractors are driven by excess of competition to make bad bargains.

But if the plans, specifications, and estimates for making a road are properly prepared, then the most safe and satisfactory way of having the work properly executed will be by letting it to a contractor.

As there is no difficulty in making an accurate estimate of the sum which a new road ought to cost, if a contractor of established reputation for skill and integrity, and possessing sufficient capital, is willing to undertake the work for the estimated sum, it will always be decidedly better to make an agreement with him than to advertise for tenders.

If a contractor cannot be got possessing the qualifications which he ought to have to justify a private arrangement, then an advertisement must be had recourse to. But when tenders are delivered in, it is very important to take care to act upon right principles in making a selection from them. The skill, integrity, and capital of the persons who make the tenders should invariably be taken into consideration, as well as the prices which they offer; for if a contractor is selected without skill, or integrity, or capital, merely because his tender is for the smallest sum, the consequence will inevitably be imperfect work, every kind of trouble and disappointment, and frequently expensive litigation.

The true principle to go upon in selecting a contractor is to lean in favour of liberal terms, and rather to overpay than underpay him. He should be made quite confident by his bargain that he will receive a fair profit for his time and labour; he will then embark in his work with spirit, and be led by a desire to gain reputation to perform his agreement to the satisfaction of all parties; but when, in following an opposite principle, a contractor is led by competition to undertake a work for a price that is too low, he starts, from the commencement, by having recourse to every species of contrivance for avoiding the fair fulfilment of what he is required to perform; every thing is done in an imperfect way; sub-contracts are made at inadequate prices; a continual contest is carried on between the contractor and the inspector, and most commonly the whole terminates in a law suit, the ruin of the contractor and his securities, and great loss to tradesmen and others by debts due by the contractor and his workmen.

V. After fixing upon a contractor, a deed of contract is to be prepared. In this the contractor should be bound to execute the work not only according to the general conditions contained in the deed, but also according to drawings and specifications to be annexed to it.

The deed should contain a clause to provide that no deviation should be made from it or the specifications, except by agreement in writing; and also a clause to provide for settling all disputes by arbitration. The other clauses which are fit to be inserted in the deed will hereafter be shown, by inserting an exact copy of a deed, according to which several portions of the Holyhead Road have been made.

VI. Before the work is commenced, an inspector should be appointed to lay out the work, to settle the levels, and to see that every particular thing required to be done is done precisely according to the specifications.

A person to be qualified to act as an inspector of a contract should have considerable experience as a civil engineer; he should be strictly sober and honest, and of reserved habits; he should avoid familiarity with those he is placed over; his disposition should be somewhat inclined to be severe, but he should be actuated at all times by the highest principles of justice and honesty in his conduct.

A chief engineer who is engaged in conducting public works will owe his success in great measure to the skill and care with which he selects the inspectors of his contracts. The necessity of making such selections forms an essential part of his occupation, and requires considerable talents to direct it.

Above all things, a chief engineer should possess

the quality of securing implicit obedience from those under him, by showing a decided superiority in the knowledge of his profession, and by acting with unsparing severity whenever the occasion may require it.

VII. It is of importance to arrange the mode of paying a contractor with punctuality; by doing so he may be kept more under control, and he will be able to perform his engagements in a more complete manner. For this purpose the deed of contract should provide that the work, as it proceeds, should be measured by the inspector every fourth week, and that a certificate should be given by him to the contractor for the value of the work that he finds executed according to the terms of the contract, deducting, in each certificate, one tenth part of the sum, to be withheld till the whole work be finished. This plan affords the best description of security for the faithful performance of a contract.

If, in place of acting upon a regular plan of paying a contractor, he is kept out of his money, he will often be placed in difficulties, and rendered incapable, however willing, to perform the conditions of his contract in a perfect manner.

As nothing can contribute more effectually to explain the proper manner of constructing a road than the deeds and drawings according to which good roads have been made, an exact copy will be now inserted of the deed of contract according to which upwards of three miles of road were made by the Parliamentary Commissioners of the Holyhead Road near Coventry. As this deed was prepared by an eminent barrister, it may be safely made use of as a precedent by road trustees, engineers, and solicitors. There will be found, in the specifications annexed to the deed, a good deal that has been already stated; but this repetition will serve to impress the more firmly those principles which should always be followed.

DEED OF CONTRACT FOR THE COVENTRY IMPROVEMENT.

Articles of agreement made and entered into the fourteenth day of November, in the year of our Lord one thousand eight hundred and twenty-seven, between Thomas Baylis, of Stratford-upon-Avon, in the county of Warwick, road-maker, of the first part; John Kershaw, of the parish of Sapworth, in the said county of Warwick, esquire, of the second part; and Alexander Milne, of Whitehall, in the county of Middlesex, esquire, on behalf of the commissioners acting in execution of an act of parliament passed in the eighth year of the reign of His present Majesty King George the Fourth, intituled, "An Act for the further improvement of the road from London to Holyhead, and of the road from London to Liverpool," to whom the said Alexander Milne is secretary, of the third part. Whereas the said commissioners having caused it to be made public that they would receive proposals for making, forming, and completing two new pieces of road, one commencing at or near the Seven Stars public house at Whitby, in the county of the city of Coventry, at the spot distinguished on the map or plan here annexed by by the letter A, and extending from thence to the present road opposite Lygon's mills, where it terminates at the spot marked on the said map or plan by the letter B; and the other new piece of road commencing in Spon Street, near St. John's Church, in the said city of Coventry, at the spot distinguished on the said map or plan by the letter C, and extending from thence to the village of Allesley in the said county of Warwick, where it terminates at the spot marked on the said map or plan by the letter D; and the several drains, walls, culverts, bridges, fences, embankments, field-gates, and works connected with the said new road, according to the specifications and general observations comprised and set forth in the first schedule hereunder written, and the plans and sections hereunto annexed, he the said Thomas Baylis delivered in the proposal contained in the second schedule hereunder written for executing the said roads, drains, walls, culverts, bridges, fences, embankments, field-gates, and works in manner aforesaid, and at the prices or sums mentioned in the same proposal; and the said proposal having been duly accepted by the said commissioners, he the said Thomas Baylis hath agreed to enter into the covenants herein-after contained for the due performance thereof respectively; and the said John Kershaw hath, at his request, agreed to become his surety in manner and to the extent herein-after mentioned; and the said Alexander Milne, on the part of said commissioners, hath agreed to enter into such stipulations as are herein-after mentioned with respect to the payment, in manner and at the time herein-after mentioned, of the consideration money to be paid to the said Thomas Baylis. Now therefore these presents witness, that in pursuance of the said proposal herein-before referred to, and in consideration of the sum of ten thousand and seventy-six pounds sixteen shillings and sixpence (the sum mentioned in the said proposal), to be paid as herein-after mentioned, he the said Thomas Baylis doth hereby, for himself, his heirs, executors, and administrators, in manner following (that is to say), that he the said Thomas Baylis, his executors or administrators, shall and will forthwith undertake, and with diligence proceed in the execution of, and duly complete or cause to be completed on or before the 25th day of March which will be in the year of our Lord one thousand eight hundred and twenty-eight, the said first-mentioned piece of road between the Seven Stars public house and Little Park Street aforesaid, and shall and will forthwith undertake, and with diligence proceed in the execution of, and duly complete or cause to be completed on or before the twenty-fifth day of March which will be in the year of our Lord one thousand eight hundred and twenty-nine, the said secondly herein-before mentioned piece of road between Spon Street and the village of Allesley aforesaid, together with all and singular the drains, walls, culverts, bridges, fences, embankments, field-gates, and other the works in the said first schedule mentioned, according to the specifications and general observations contained in the same schedule, and the plans and sections hereto annexed, and at the price or sum mentioned in the said proposal contained in the said second schedule; and further, that the said Thomas Baylis shall and will, in all respects, faithfully, readily, and diligently abide by, observe, perform, and keep the said several terms and stipulations mentioned in the said specifications and general observations, and pursue the said plans and sections according to the true intent and meaning thereof, and shall not nor will make any deviation therefrom without the licence, consent, and approbation of the engineer for the time being of the said commissioners, in writing, for that purpose first had and obtained; and that the said Thomas Baylis, or some experienced and responsible person on his part, shall from time to time pay due personal attendance to the said several works, as the same shall be carrying on, and shall at proper times be ready to be conferred with by the inspector appointed by the said commissioners; and these presents further witness, that in pursuance of the said agreement on this behalf, and in consideration of the premises, he the said Alexander Milne doth hereby for himself, his executors and administrators, covenant, promise, and agree with and to the said Thomas Baylis, his executors, administrators, and assigns, in manner following; (that is to say,) that in case he the said Thomas Baylis shall and do well and truly perform and keep the several stipulations and agreements herein contained, referred on the part of him the said Thomas Baylis, his executors or administrators, to be performed and kept according to the true intent and meaning of these presents, then they the said commissioners for the time being, or such person or persons as shall from time to time be appointed by them for that purpose, shall and will well and truly pay or cause to be paid unto the said Thomas Baylis, his executors, administrators, or assigns, the said sum of ten thousand and seventy-six pounds ten shillings, in manner hereinafter mentioned; (that is to say,) at the end of each twenty-eight days from the time when the making of the said roads or works shall be begun upon as aforesaid, and shall pay or cause to be paid to the said Thomas Baylis, his administrators, executors, or assigns, nine tenth parts of the sum which shall be payable in respect of or be taken as the consideration for the work that shall have been done in the course of the twenty-eight days immediately preceding; and further, that for the purpose of ascertaining how much of the said sum of ten thousand and seventy-six pounds ten shillings shall be taken for the consideration for each twenty-eight days work, the engineer for the time being of said commissioners, or his assistants, shall, at the end of each twenty-eight days, measure the work done, or otherwise estimate the same, and shall report or state how much of the said sum of ten thousand and seventy-six pounds ten shillings ought to be taken as the consideration for the work done during the twenty-eight days immediately preceding each such admeasurement or other estimate, and that the nine tenth parts of such stated or reported price or consideration shall, immediately after each such admeasurement or other estimate, be paid to the said Thomas Baylis, his executors, administrators, or assigns as aforesaid* ; and further, that within one month after the whole of the said roads, drains, walls, culverts, bridges, fences, embankments, field-gates, and works shall be completed, according to the true intent and meaning of these presents, all the residues which shall then remain unpaid of the said sum of ten thousand and seventy-six pounds ten shillings shall be paid to the said Thomas Baylis, his executors, administrators, or assigns; and the said John Kershaw, in pursuance of the said agreement in this behalf, doth hereby bind himself, his heirs, executors, and administrators, unto the said Alexander Milne, his executors, administrators, and assigns, in the sum of twenty thousand pounds, to be paid to the said Alexander Milne, his executors, administrators, or assigns, by the said John Kershaw, his heirs, executors, or administrators† : Provided, and it is hereby agreed and declared, that if he the said Thomas Baylis, his executors and administrators, shall duly observe, perform, and keep the cove-

* A clause should be added to provide that when each monthly measurement and estimate are made, the contractor should declare what claims he may have for any extra work.

† In subsequent contracts less security was required, the stopping of the one tenth of each month's payment being found the best security.

nants and agreements herein contained on his and their parts, then the above written obligation on the part of the said John Kershaw shall be void, or otherwise shall remain in full force; and it is hereby further covenanted, concluded, and agreed, that in case any addition, alteration, or deviation shall be made at the desire of the said engineer, in writing, to, in, or from the plans or particulars, or either of them, every such addition, alteration, or deviation shall be paid for separately, and shall be subject to the like agreement, with respect to materials and workmanship, as herein-before contained with respect to the making and completion of the roads and works, according to the aforesaid plans and particulars, and as if such additions, alterations, and deviations had formed part of and been specified in such plans or particulars; and that no such addition, alteration, or deviation shall be considered as vitiating this present agreement, or shall in any wise affect the same, except so far as is mentioned in the stipulation last herein-before contained; and if such alterations shall increase or lessen the works herein-before stipulated to be performed, the same shall be paid for or deducted, as the case may be, at rates corresponding with the value of similar works herein-before contracted for: Provided, and it is hereby further covenanted, concluded, and agreed upon, by and between the said parties to these presents, that in case any doubt, dispute, or question shall arise, touching and by virtue of this agreement, or touching or concerning the true intent and meaning of these presents, all such doubts, disputes, or questions shall be referred to the decision of two referees, one to be named and appointed by the said Alexander Milne on behalf of the said commissioners, and the other to be named and appointed by and on the behalf of the said Thomas Baylis; and in case such two persons shall not agree in the premises, and make their decision and determination in regard thereto in writing, under their hands, in one month after the same shall be referred to them, then, and in such case, all such doubts, disputes, or questions shall be referred to such third person as the two persons so to be chosen aforesaid shall for that purpose nominate, and whose decision and determination thereon shall be binding and conclusive between the said parties, so as the same may be made in writing within one calendar month from the time the same shall be so referred to him as aforesaid. In witness whereof the said parties to these presents have hereunto set their hands and seals, the day and year first above written.

The first Schedule referred to in and by the foregoing Agreement.

HOLYHEAD ROAD.

Coventry Improvement Specification.

The first improvement commences on the present road, near the Seven Stars public house, at the point marked A on the plan and section (Plate VI.

fig. 1.)*: having ascended the small hill in Mr. Troughton's land, it crosses the Folly Lane, along which it runs a short way, and descends to the river, which is to be passed by a bridge of thirty feet span†; it then ascends the opposite hill, passing near the mill in the hollow; and, after crossing a garden and some meadows, terminates on the present road, in the common, at the point marked B on the plan and section.

The second improvement commences in Spon Street, near Saint John's Church, at the point marked C in the plan and section (Plate VI. fig. 2), and after passing over the site of two houses and their premises, runs in front of the weavers' row of houses, and crossing the Chapel Lane, goes over the Windmill Hill, and continuing through the fields in almost a straight line, passes to the south of the city reservoir, over a stream; and from thence, over the hill in front of Mr. Wall's house, to the widow Hewett's house, and entering on the present road continues over the bridge, and finally terminates in the village of Allesley, at the point marked D on the plan and section.

ROAD-MAKING.

The dotted lines on the sections represent the natural surface of the ground in the longitudinal

* The plans in this plate are on a reduced scale. For working drawings the longitudinal scale should be three chains to an inch, and the vertical scale thirty feet to an inch.

† The specification for building this bridge is in Chapter VII., on Road Masonry.

direction of the new lines, about the middle of the space to be occupied by the road. The black lines represent the finished longitudinal surface of the foundation for the road materials. The figures above the vertical lines denote the depths of cuttings and heights of embankments. The rates of inclination which are on the first improvement are as follows:—1 in 30, 1 in 40, 1 in 36, 1 in 60, 1 in 100, 1 in 56, 1 in 35; and on the second, 1 in 35, 1 in 66, 1 in 51, 1 in 253, 1 in 180, 1 in 40, 1 in 101, 1 in 41, 1 in 40, and 1 in 30*; but it is expressly stipulated, that the contractor is to satisfy himself by his own measurements, or in any way he may think proper, as to these depths and heights; and all irregularities of the surface of the ground to be cut down or embanked upon, as no extra claim will be admitted.

The breadth of the finished road is to be 35 feet; 30 feet for the carriage-way, and 5 feet for the footpath. (Plate VII. fig. 1.)

The slopes of all embankments from the outsides of the finished road are to be two horizontal to one perpendicular, neatly dressed, and covered with green sod, at least four inches thick, evenly laid and closely jointed. (Plate VII. fig. 2.)

The slopes of the cuttings on the southern side to be three horizontal to one perpendicular, those of the northern side to be two horizontal to one perpendicular; these slopes are also to be covered with sod, the green side placed uppermost, evenly

* These inclinations are shown in the working drawings.

laid and properly jointed; it may be procured on the space to be occupied by the road and its side slopes. (Plate VII. fig. 3.)

The surface of the foundation for the materials of the carriage road is to be formed level from side to side. The surface of the foundation for the hard materials of the footpath is also to be level, and to be six inches above that of the carriage road. In the cuttings, the breadth between the bottoms of the side slopes is to be thirty-three feet. (Plate VII. fig. 1.)

Upon the level space prepared for the road materials, a bottom course, or layer of stone, is to be set by hand in form of a close firm pavement. (Plate III. fig. 3.)

The stones set in the middle of the road to be seven inches in depth; at nine feet from the centre five inches; at twelve feet four inches; and at fifteen feet three inches; and to have a curving surface; they are to be laid on their broadest edges lengthwise across the road, and the breadth of the upper edge is not to exceed four inches in any case. All the inequalities of the upper part of said pavement to be broken off with a hammer, and all the interstices to be filled with stone chips firmly wedged or packed, so as to form a convexity of four inches in the breadth of fifteen feet from the centre. The stones used for this purpose are to be such as will not waste by exposure to the weather.

The bed of pavement or set stone is to be covered with a layer of Nuneaton stone, to be approved of by the inspector; it is to be laid six inches thick on the middle of the road, eighteen feet wide. These stones are to be broken into pieces as nearly cubical as possible; the longest and largest piece to go through a circular ring of two inches and a half in the inside diameter.

The stones must be broken off the road, and riddled on a sieve of one inch square meshes, and approved of by the engineer or his assistant before they are put on the road. The shoulders or sides of the road are to be covered with the small stones which pass through the aforesaid meshes, properly cleansed and selected, or with good sharp coarse gravel, well cleaned and separated from all earth, and having no pebbles larger than one inch and a half in diameter, to be approved of by the engineer or his assistant. This course of gravel is to be seven inches and a half thick at nine feet from the centre; six inches and a half at twelve feet from the centre; and two inches and a half at the sides, or fifteen feet from the centre.*

The Nuneaton stone is to be covered with one inch and a half in depth of good binding gravel, to be laid on at such times and in such way as may be directed; and the whole surface, from side to side, to be kept properly dressed and levelled, until the whole work is certified to be completely finished.

The footpath is to be coated with six inches of sandstone, broken to the same size as stated for the

* Since this specification was drawn up, it has been considered better to increase the depth of pitching at the sides, so as give a convexity to the road surface of six inches instead of nine.

road materials, and is to be covered with gravel, so that the surface of the footpath shall be on a level with the middle of the road.

DRAINAGE.

Along the outside of the carriage road (all embankments excepted), drains are to be cut, ten inches wide at bottom, fourteen inches at top, and ten inches deep below the surface of the bed for the road materials. (Plate VII. fig. 4.) Mitre drains are to be made from the middle of the road into these drains, forming such angles at the middle as may give a declivity for conveying the water into the side drains; they are to be nine inches wide at bottom, twelve inches at top, and ten inches deep; there are to be thirty of these drains per mile (embankments excepted). These drains are to be filled with rubble stone, connecting with the bottom course of road materials. An edging of turf, with the green side out, not less than six inches in depth and five inches in thickness, is to be set in a neat compact manner on the road edge of the footpath, and the top thereof is to be covered with the binding gravel. On the outside of the footpath, along the border for the thorns, a turf six inches high and four inches thick is to be set in the same way over the whole line, excepting on embankments.

There are to be eight cross drains of dry stone masonry, each eighteen inches in the clear, constructed in every mile of the road. These cross drains are to be continued under the fences into the ditches on each side of the road. The side walls to be sixteen inches thick, faced on both sides, eighteen inches high on the upper end, and twenty-three at the lower. The top of the walls to be level across the road, and the bottom to have an inclination of one inch in every ten feet. The stones at top, on which the covers are to be laid, are to project about two inches and a half into the open space on each side, leaving about thirteen inches clear between them. The covers to be sound stone, not less than four inches thick, and twenty-seven inches long. They are to be neatly jointed, closely laid together, and properly bonded on the side walls, and covered with four inches of turf. A concave pavement, with stones not less than five inches deep, to be laid between the side walls, as shown in the drawings (Plate IV. fig. 5). The whole of the building to be placed so low as to admit six inches of mould to be laid between the covers and the bottom course of stone, without raising the longitudinal surface of the road. When the cross drains are under an embankment, the same are to be carried to the extremities of the bottoms of the slopes. Should any drains of a different size be wanted, their situation, number, size, and value to be determined by the inspector. The water from the surface of the road to be introduced into the cross drains by as many side openings or inlets as there are cross drains, ten by sixteen inches, built on each side, and covered with stones at least twenty-six inches long, fourteen inches broad, and not less than two inches and a half thick. The bottom of said covers to be five inches above the side drain, and the whole of each opening to be on the outside of the driving way. The ends of the cross drains to be secured by strong pavements, four feet three inches by two feet three inches. The water collected in the side drains of the road to be introduced into the cross drains by a row of paving-stones across the course, so raised as to prevent it from passing the opening of the cross drain; and the outer row of paving-stones below the discharging end to be large stones, sunk so deep as to secure the whole from being injured by the current of water. The lower ends of the drains to be secured by wing walls, at least five feet in length, and the same at the upper end, where they are connected with a watercourse, and to be covered with two rows of swarded turf; the lower one with the swarded side down, the other with the swarded side up. Wherever springs are found in the surface of the ground on which the road is to be made, or in the cuttings, drains to be made the same as on the outside of the carriage road before described, for carrying the water into the ditches or natural watercourses, by proper under-draining. Open cuts to be made whenever they are necessary for carrying off the water from the ditches into the natural watercourses; and these drains to be two feet wide at the top, ten inches wide at the bottom, and eighteen inches deep. Through all the cuttings on the footpath side of the road, a drain, one foot square, is to be made along the lower edge of the slope, and filled with rubble stone. This drain is to be made between the footpath and the bottom of the slope; the bottom of it to be eighteen inches below the upper surface of the finished footpath, and at the opposite side of the road, the bottom of it is to be one foot below the under surface of the metal pavement, as shown in fig. 4. An open catch drain is to be made through the cuttings, above the quicksets, and also at the tops of the slopes, where the ground inclines to the road; to be one foot deep, sloping three to one on the field side, and one to one on the road side, meeting in an angle at the bottom, and the whole neatly dressed and covered with green sod.

FENCING.

The fencing to be constructed as shown in the section; a ditch to be cut, and a mound to be raised, together occupying eight feet; the ditch to be on the field side, the mound to be cut out of the natural ground, four feet wide at top, ten inches wide at bottom, and two feet and a half deep.* The mound, of four feet wide, is to be raised by a sod, with the green or swarded side out, to the height of fourteen inches above the side channels of the road, and the top to be rounded from the ditch to the top of the sod. Two rows of quicksets to be planted on the ditch side of the mound, as shown in the section. Nine plants to be set in

* In wet land the drain should be at least four feet deep.

each lineal yard; they are to have good roots, to be two years transplanted, to be put in between the first day of November and the last day of March. A trench, eighteen inches wide and fifteen inches deep, is to be cut, and filled with good vegetable mould, in the middle of which the two rows of quicksets are to be planted: in all the cuttings the quicksets are to be planted at the distance of eighteen inches from the footpath on the one side, and the carriage road on the other side. (Plate VII. fig. 4.) On all embankments they are to be planted so that the distance between the middle of the rows on each side shall be thirty-eight feet. (Plate VII. fig. 5.) Particular attention is to be paid to the preparation and quality of the earth of the quickbed, and every thing connected with the planting of the quicksets.

The quicksets are to be protected by two rows of posts and rails on each side of the road, three rails in each length; the posts are to be five feet long, and at least five by three inches of good oak; the rails to be not more than eight feet long, and three inches and a half by two inches and a half, and may be of good elm, ash, or fir timber. In each length of rail a prick post is to be driven into the ground; they are to be placed in the middle between the posts, to be at least twelve inches in the ground, and well fitted, and strongly nailed to each rail. (Plate VII. fig. 6.) A mound, two feet wide and fourteen inches high, to be made below the railing placed on the top of the embankments. Each field must be fenced off with the posts and rails before any part of the road work is commenced in that field, or before any of the hedges or ditches now existing on the lands be removed or touched. Ten field gates, with iron hinges and fastenings, and ground posts, all similar to the best kinds used in the neighbourhood, to be furnished and erected; and should a greater or less number be required, they are to be allowed or deducted from the contract at so much per gate. At each gate, drains with good draining tiles not less than ten inches, or of brick one foot wide, to be laid in the sides of the road; and drains of the same construction as the cross ones, one foot square in the clear, to be made in the field or outside ditches: the length of these drains to be twelve feet, and a road to be made over them into the fields, eight feet in breadth, and covered with broken stone of the same quality as used for pitching the road; to be eight inches deep, and extending into the fields ten feet at least beyond the line of the quicksets. No inclination from the road into the fields to be more than one in sixteen, and all the gates to open into the fields.

DEPÔTS FOR HOLDING REPAIR MATERIALS.

Eight depôts to be erected in each mile of road, in such places as may be pointed out by the engineer or his assistant. (Plate IV. fig. 7.) They are to be built of stone and lime, twelve yards long, three feet high above the side channel of the road, and to be founded as low as necessary below that, to give stability to the work. The ends to be two yards and a half in the clear at the bottom, and to rise one yard and a half at top. The thickness of the work to be eighteen inches throughout for the height of three feet; the work under that to be two feet thick. The top of the back, side, and slopes to be coped with large stones set on edge, and even, and flagged with sandstone in the rough, neatly jointed and well bedded.

The back and ends of the depôts are to have a mound of earth thrown up against them, eighteen inches high on the outside, and eighteen inches or two feet on the base, rounded off on the top, and faced with sod if necessary, and the regular quantity of quicksets planted in it, which are to be protected by the field row of posts and rails described before; a tile drain to be laid in front of the open, thirteen yards long, and ten inches wide.*

GENERAL OBSERVATIONS.

All the lines to be marked out by the chief engineer to the parliamentary commissioners, or his assistant, and the general formation of the road is to be to his satisfaction; he is also to be satisfied with the solidity of all embankments before the foundation or bottom course of pavement is laid on them. The stone used for the pavement, and

* A specification in the original schedule for building a bridge has been transferred to Chapter VII., on Road Masonry. The specifications also in that chapter for depôts and inlets were taken from this schedule.

the packing and setting of the same, are in all cases to be approved of by him before any broken metal is laid on. He is also to be satisfied that the top metal is of proper quality and dimensions before the binding is laid on, and that the cross drains are properly constructed and firmly backed before earth or turf is placed on them. A passage is to be constructed from the embankment at the mill near the bridge, to admit carts to Barnewall's mill and premises: it is to be fourteen feet wide, to have an inclination not more than one in sixteen, to be covered with broken stone ten inches thick at the middle, and six inches at the sides; a dry stone wall to be built on the outside, to retain the earth; to have a batter of two inches in three feet, and to rise three feet above the surface of the finished road, as a protecting parapet; to be fifteen inches thick at top, and increasing in thickness by an offset of three inches on the inside for every foot to the foundation; the top course to be set in good lime and mortar. A paved channel one foot and a half wide is to be laid along both sides of the road, from top to bottom; where the road crosses Folly Lane there will be a cutting of two feet. The lane must be lowered at each side of the road, and properly levelled to an inclination of one in sixteen for the whole width of the road. The part thus broken up must be coated with six inches of broken sandstone of the best quality in the neighbourhood. Tile drains are to be laid along the side of the road for the whole width of the lane at the point of intersection.

Where the road crosses the footpath at the upper angle of the weavers' row of houses on the Allesley land, a cutting is to be made of two feet six inches; the footpath must be covered on each side, and dressed with gravel, so as to give a good and commodious passage to and from the road. Where the road crosses the Chapel Lane there is to be four feet of filling; the lane must be embanked on each side to the height of the road, and have inclinations each way of one in sixteen; the side slopes are to be two horizontal to one perpendicular, and the top surface when finished to be twenty-one feet wide; both these side roads or lanes are to be fenced in the same manner as the main line of road, and to be coated with eight inches of broken sandstone for the width of eighteen feet, and with gravel for the width of eighteen inches on each side, to be eight inches deep where it joins the broken stone and six inches at the sides.

Where the road crosses the lane between Mr. Booth's and Mr. Carter's land, a filling is to be made of two feet nine inches. The same must be embanked up to the road on each side; to have inclinations each way of one in sixteen, and to be twenty-one feet wide on the surface, with slopes of two horizontal to one perpendicular; to be covered with five inches thick of pebbles, eighteen feet wide, and gravelled eighteen inches at each side; the gravel to be three inches thick next the broken stone, and six at the sides; no fencing will be necessary on these side roads.

The second Schedule referred to in and by the foregoing Agreement.

Sir,

I hereby engage to execute the works of the proposed improvement of the Holyhead Road at Coventry, according to the plan, section, and specification, for the sum of ten thousand and seventy-six pounds sixteen shillings and sixpence.

I am, Sir,

Your obedient servant,

THOMAS BAYLIS.

	£	s.	d.
First lot - - -	4,140	0	0
Second lot - -	5,936	16	6
Total - -	10,076	16	6

To Alexander Milne, Esq.,
1, Whitehall, London.

Signed, sealed, and delivered by the above-named Thomas Baylis, in the presence of

John Macneill. THOMAS BAYLIS.

Signed, sealed, and delivered by the above-named John Kershaw, in the presence of

John Macneill. JOHN KERSHAW.

Signed, sealed, and delivered by the above-named Alexander Milne, in the presence of

John Macneill. ALEXANDER MILNE.

CHAPTER IX.

IMPROVING OLD ROADS.

MR. TELFORD gives the following account of the state of the turnpike roads in 1819, in his evidence before the committee of the House of Commons on the highways of the kingdom: — "With regard to the roads of England and Wales, they are in general very defective, both as to their direction and inclinations; they are frequently carried over hills, which might be avoided by passing along the adjacent valleys; the shape or cross sections and drainage of the roads are quite as defective as the general directions and inclinations; there has been no attention paid to constructing good and solid foundations; the materials, whether consisting of gravel or stones, have seldom been sufficiently selected and arranged; and they lie so promiscuously upon the roads as to render it inconvenient to travel upon them, and to promote their speedy destruction. The shape of the road or cross section of the surface is frequently hollow in the middle; the sides encumbered with great banks of road dirt, which have accumulated in some places to the height of six, seven, and eight feet; these prevent the water from falling into the side drains, and also throw a considerable shade upon the road, and are great and unpardonable nuisances. The materials, instead of being cleaned of the mud and soil with which they are mixed in their native state, are laid promiscuously on the road: this in the first place creates an unnecessary expense of carriage of soil to the road, and afterwards nearly as much in removing it, besides inconvenience and obstruction to travelling."

The committee of 1819, by attributing in their report the imperfect state of the roads to the negligent and culpable conduct of the trustees who had the management of them, roused the attention of the public to the subject, and thus led to the introduction of an improved system of management. But although a considerable change for the better has taken place since 1819, many of the defects described by Mr. Telford still remain; and all that has been done towards their removal falls far short of what ought to have been done to put the turnpike roads into complete order.

In improving old roads, nearly the same objects should be attended to as in making new ones; such, for instance, as the direction, the longitudinal inclinations, the breadth, form, and hardness of the surface, the drainage, and the fencing.

For the purpose of ascertaining in what respect an old road is complete or defective in these points, the following queries have been prepared. The answers that can be given to them will at once show what is the state of a road.

1st. Is the direction of the road in the shortest line that can be found without having to pass over steep hills or other obstacles?

2d. What are the rates of inclination on the hills? In ascending towards a height, or ridge of country, that must be crossed, has the road one gradual ascent, or has it one or more descents, thus making two or more hills instead of one hill?

3d. What is the breadth of the road? Is it every where exactly the same? Is it defined by side channels having along them curb stones or borders of grass sods?

4th. Are the channels on each side of the road on the same level? Is the convexity of the surface uniformly the same in every part along the whole length of the road?

5th. Is there a footpath? What is the height of it above the side of the road? What is its breadth? Of what materials is it composed?

6th. Is there any waste land between the road and the fences of the road? In what state is it?

7th. Is the surface of the road higher than that of the adjacent fields?

8th. Of what materials does the crust of the road consist? What is the depth of them in the centre of the road, at a distance of five feet on each side of the centre and at the sides?

9th. Are there sufficient drains for carrying off all rain and other water?

10th. Are the fences low? Are they raised on ground of the same level on both sides of the road? Are they of the same height on both sides, and parallel to each other?

The answers which can be given to these queries will show what the defects are of any road to which they are applied, and what is requisite to be done to improve it.

With respect to the turnpike roads as they now are, it will be found upon an inspection of them that, in regard to their direction, they are universally defective. Scarcely any road between two places is in the best line with respect to distance and hills. The reason of this is that the present lines are the same, except those made of late years, as they were when first established by the aboriginal inhabitants of the country as footways or horse-tracks. Let a map be made of the road from London to Edinburgh, to Carlisle, to Liverpool, or to any distant town, and this fact will be fully sustained.

The first step which should be taken towards the improvement of the principal roads of the kingdom is to make surveys of the mail coach roads: this work should be done by government. The engineers employed should also be required to make plans and estimates for the improvements which may appear necessary; and the trustees of every principal road should be furnished with copies of the surveys, and of the plan and estimates for improvements relating to the road under their care.

The number of single mail coach miles daily travelled in Great Britain, including pair horse coaches, is 17,549.* The expense attending the

* Seventh Report of the Commissioners of Post Office Inquiry, p. 4.

surveying of them should not exceed 3*l.* a mile; so that the whole expense to be incurred on this important preliminary step for the improvement of these roads would not be of a large amount.†

Whenever the improvement to be made on an old road does not require a departure from the present line, the road should first be put into a proper form, as respects breadth and convexity, according to the rules already laid down. A paved foundation should be made from 12 to 30 feet wide, according to the funds of the road, and a coating of broken stones six inches thick should be laid on; a regular footpath should be made; all the old high and crooked fences should be removed, and low ones substituted in their place, parallel to each other, and at a proper distance from the road; and particular care should be taken to provide a sufficient number of drains.

Where the old road is below the level of the adjoining fields, it should be raised by embanking, so as to be at least two feet above them.

If it is not considered advisable to remove the old fences, where the space between them is wider than is necessary for the roadway and footpath, the surplus portion, or waste, should be put into order; for no road can have a finished appearance unless the whole space between the fences is arranged so as to have a regular and uniform shape. This operation will also assist very

* This subject will be again referred to, in Chapter XII., on Road Legislation.

much in contributing to the dryness and preservation of the road. On this point Mr. Telford makes the following observations in his third annual report on the Holyhead Road: —

" I cannot too often repeat, that a surveyor should not feel satisfied that he has done his duty until the whole breadth of ground belonging to a road between the fences is put into perfect order, as this shows skill, attention, and good workmanship. A certain space, say six feet, should be formed into a footpath of one regular breadth, with a surface made with a coating of strong gravel, or small broken stones, at least six inches deep; thirty feet should be allotted to the roadway, to be formed of one regular convexity, with the use of a properly shaped level*; one side channel should be formed by the sod margin of the footpath abutting on the side of the road, and the other by the sod margin of a flat mound of earth, of the same form as the footpath; and the whole waste between the fences should be filled or levelled, so as to have a perfectly smooth surface. The wastes should also be sown with grass-seeds; and where the soil is clay, the scrapings of the road should be carefully spread over them, till they become firm. When the fence of a road is a hedge, this should be cut and clipped every year by the surveyor, at the expense of the trustees; and the work should be done in such a manner as to leave the side and horizontal lines of the hedge perfectly straight and even.

* See Plate VII. fig. 8.

"In order to assist the surveyors in putting their roads into a good shape, I have drawn up the following specification: —

"*Specification for the Regulation of the Surface between the Fences, so as to establish uniformity in the Cross Section.*

"1. The road is to be thirty feet wide, exclusive of footpaths, with a fall of six inches from the centre to the side channels.

"2. A sod to be laid on each side of the road, eight inches wide, and six inches in thickness, and in such a manner as to form a sloping edge; the top surface of the sods on each side to be exactly on the same level.

"3. On one side of the road a footpath to be made behind the sod; it is to be six feet wide, and to have an inclined surface of one inch in a yard towards the road; and another sod to be laid along the outer edge of the footpath, eight inches wide, the top of it on a level with the footpath.

"4. On the other side of the road a flat mound of earth is to be formed behind the sod, on a level with the top of it, six feet wide; the surface of this mound is to be sown with grass seeds.

"5. The waste land on each side, where there is any, between the footpath, or the mound, and the road fences, to be dug over and made quite smooth; when these wastes are covered with grass, the sod to be pared off each breadth, and laid on the breadth last dug; when they are not in grass, the new surface is to be sown with grass seeds.

"6. If there is a ditch on the road side of the fence, or if the road fence consists of a high bank, a new post and rail fence is to be made close along the footpath or mound, with a ditch on the field side, at least three feet deep."

If the foregoing rules were strictly attended to, the safety of fast travelling by night coaches would be very much increased. The accidents which occur by night arise chiefly from coachmen getting off the road, and running the wheels of coaches on high footpaths or other high banks of earth immediately on the sides of the road; but if no footpath were higher than six inches above the side channel of the road, and if a flat mound were formed of the same height on the side opposite to the footpath, coachmen, on getting off the road in fogs or snow storms, would be able to pull into it, or stop, without any danger of being overturned.

The parish roads might be much improved by attention to a few general rules. Twenty feet in breadth should be carefully set out and defined by a row of sods on each side.

The surface should be brought to a convexity of six inches from the centre to the sides, by laying on good road materials. The ruts should be filled with hard materials from time to time.

The space on each side between the sods and the fences should be made smooth, with an inclination of one inch in a yard from the road to the fences. Drains should be made along the fences, and all watercourses and drains connected with the road should be constantly kept open, and free from weeds.

Those parish roads which are very narrow, and whose surface is below the level of the adjoining fields, and on which streams of water are constantly running, should be new made, by raising them with earth, and forming a roadway of good materials on the embankment.

CHAPTER X.

REPAIRING ROADS.

THE business of repairing a road should always be managed on a regular and fixed plan.

The following matters require particular attention:—

1st. The quality of materials.

2d. The quantity to be put on per mile per annum.

3d. The preparation of the materials.

4th. The method of putting them on the road.

5th. The number of labourers to be employed.

1st. With respect to the quality of the materials to be used: the hardest should always be preferred; for it should ever be borne in mind that hard stones brought from a distance are found by experience to be cheaper in the end than those of a softer kind which may be got near the road at a much lower price.

Another reason for making use of the hardest materials that can be procured is the greatly increased labour of horses, which is occasioned by working into a smooth surface often renewed coatings of weak materials. With respect to the subject generally of road materials, it may be observed, that the best descriptions consist of basalt, granite, quartz, syenite, and porphyry rocks.* The whinstones found in different parts of the United Kingdom, Guernsey granite, Mountsorrel, and Hartshill stone of Leicestershire, and the pebbles of Shropshire, Staffordshire, and Warwickshire, are among the best of the stones now commonly in use. The schistus stones will make smooth roads, being of a slaty and argillaceous structure, but are rapidly destroyed when wet, by the pressure of wheels, and occasion great expense in scraping, and constantly laying on new coatings. Limestone is defective in the same respect. It wears rapidly away when wet, and therefore, when the traffic is very great, it is an expensive material. Sandstone is much too weak for the surface of a road; it will never make a hard one, but it is very well adapted to the purpose of a foundation pavement. Flints vary very much in quality as a road material. The hardest of them are nearly as good as the best limestone, but the softer kinds are quickly crushed by the wheels of carriages, and make heavy and dirty roads. Gravel, when it consists of pebbles of the hard sorts of stones, is a good material, particularly when the pebbles are so large as to admit of their being broken; but when it consists of limestone, sandstone, or flint, it is a very bad one; for it wears so rapidly that the crust of a road made with it always consists of a large portion of the earthy matter to which it is reduced. This

* For the hardness of some particular kinds of stone, see Appendix, No. III.

prevents the gravel from becoming consolidated, and renders a road made with it extremely defective with respect to that perfect hardness which it ought to have.*

* The following is a description of an experiment that has been made of the use of iron materials, in the shape of small cubes, mixed with broken stones:—

Road made partly with broken stone and partly with pieces of cast metal, laid on over a sub-pavement of rubble stone.

This plan has been lately tried on a part of the Holyhead Road between London and Birmingham, and appears to possess some important advantages, which, however, can only be ascertained with certainty by a trial on a more extended scale, and in a situation where there is a much greater thoroughfare of horses and carriages.

The plan simply consists in laying on pieces of cast metal on the surface of a road, previously well constructed in the manner recommended by Mr. Telford, and adopted in all the new works on the Holyhead Road; that is, with a rubble pavement of large stones, covered with six inches of good hard broken stone. This layer of stone should be laid on in two successive operations; the first coat should not be more than three inches thick, and which should be kept constantly raked until it became nearly consolidated; the second coat should then be applied, and raked in very carefully the moment a carriage track appeared in it. When this coating becomes consolidated, so that a carriage could pass over it without leaving a rut, the iron may be applied. This may be done by a labourer. He should be provided with an iron instrument similar to a marline-spike, and about twelve inches long; the upper six inches should be round, and the lower six should be tapered down to a square point. With this tool and a mallet a hole should be made in the road large enough to receive the cube of metal, which should be stuck firmly down, until the upper surface of the iron is on a level with the surface of the road. The stone chip caused by making the hole should then be packed round the iron, and

2dly. With respect to the quantity of materials to be put on a road in the course of a year: this should be regulated by the traffic and the durability of the materials. The object to be secured is, to give the road a degree of strength sufficient to make it smooth and hard, not only in ordinary wea-

---

beat down by the mallet. One of these iron cubes should be placed in the manner above described in every four square inches of surface, and the road, if the work be done in summer, should be frequently and copiously watered, until the iron cubes become perfectly fixed and firm in their places, which will be the case in three or four days, even with a limited traffic over the road.

It was in this way that the experiment was made on the Holyhead Road; the cubes of iron were put in in the month of March 1835, and became consolidated in about three days: very few were thrown out of their places by the horses' feet or the wheels of carriages during that period; and after they became consolidated there was not one observed to move. They have been now on the road two years, and they appear to have suffered scarcely any wear. The road has remained in a very perfect state up to the present time; and during the whole of the last winter, which was singularly unfavourable for roads, it has required scarcely any scraping, and no repairs whatever from the first time it was laid down. There is nothing in the appearance of the road to indicate that it is at all different from common stone roads, and it requires a close examination to detect the iron. The horses do not slip in the least, and the wheels of carriages pass over it apparently with great ease of draught.

In streets iron may prove of great advantage, as it will certainly diminish to a great extent the nuisance of constantly picking up the surface and laying on new materials. The iron of which the cubes are cast may be of the very worst and cheapest description, and will probably not cost the third part of common castings.—J. Macneill.

ther, but during the whole of the winter months. The materials should be quarried, carted, and broken by contract. When brought to the road, they should be packed in depôts, or laid up on the wastes, in regular shaped heaps, so as not to interfere with the side channels of the road.

3dly. When the materials are stone they should be broken, as before described, to a size of a cubical form, not exceeding two inches and a half in their largest dimensions.

When gravel is used, the persons who dig it should be required to pass it through sieves before it is carted, so that no gravel pebble less than one quarter of an inch in diameter should be carried from the pits to the road; and when there it should be again sifted by the labourers, so as to separate the pebbles that are less than three quarters of an inch in diameter from the rest; all the pebbles exceeding one inch in diameter should be broken.

4thly. The materials, after they have been properly prepared, should be laid on in small quantities at a time.

In those places where the surface of the road has become much worn, a coating of two inches and a half of materials should be laid on; that is to say, a coating only a single stone in thickness, when stones are used; and when gravel is used, a coating not exceeding one inch in thickness should be laid on. If more materials are necessary, they should be laid on in successive coatings after the first coatings are worked in. After a coating has been laid on, the edges of it should be covered with the scrapings of the road to the breadth of eighteen inches. This will contribute very much to relieve the horses when drawing carriages over it, and to its being quickly consolidated.

The work of repairing roads by laying on new coatings of materials ought to be done between the months of October and April, and when the surface is wet. By laying on the materials at this season of the year in thin coatings, they are soon worked into the surface, without being crushed into powder, and without producing any great distress to horses drawing carriages over them.

Care should be taken to lay on small quantities of material, even so little as a shovelfull, on the appearance of a rut or hollow. The practice, which is in some places followed, of picking up and loosening the surface before laying on a new coating of materials, destroys a great quantity of the old ones, is attended with a heavy expense, and is productive of no good whatever. However hard the surface of a road may be, when a coating is laid upon it, this coating keeps the surface damp, and softens it, so as to let the new stones fasten themselves into it.

5thly. When the funds will admit of it, a road should be divided into districts of four miles each; and a foreman, with three labourers, should be appointed for each district. The foreman and one or more of the labourers should be daily on the road, taking care that the surface and side channels are kept clean, and making good any defect as soon as it appears.

The foreman should work with the men: he should take care that the orders of the surveyor are attended to, and be able to measure road work. When the men are not wanted on the road, they should be employed by task work in getting and preparing materials.

A regular plan should be arranged, and strictly adhered to, for keeping the water channels and drains always open and free from dirt.

In the month of October in each year, every water channel and drain should undergo a general repair, and be cleared of all deposited earth and weeds.

At the same time all ruts and hollows should be carefully filled with materials, and all weak parts of the surface coated with materials; that is to say, the road should be put in every respect into a complete state of repair, so as to preserve it from being broken up during the approaching winter.*

Nothing is more important to be attended to, in order to preserve a road in good order, than the continual scraping of it. This work should be done after every heavy fall of rain, so as never to

* M. Berthault Ducraux, Ingenieur des Ponts et Chaussées, in his Treatise " De l'Entretien des Routes," implies, from the expression of " all ruts " here made use of, that the principal roads of England are cut into ruts. As the rule here recommended to be followed applies to all roads, and as there are many, but not of any general use, still rutted, the expression is a correct one, although none of the principal roads have ruts on them. The statement of M. Navier, in his work on roads, on this point, which is quoted by M. Berthault Ducraux, is quite correct.

allow the mud to be more than half an inch thick. Throughout the winter every road, where the traffic is great, should be scraped once at least a week. By doing this the surface becomes dry in the intervals between showers; ruts are not formed, and the workmen, while scraping, discover the parts of the surface which require materials to be laid on in order to prevent hollows and holes from being made.

The great expense necessary to be incurred in scraping, when the materials consist of gravel or stones of inferior quality, points out the expediency of taking pains to procure the hardest stones. Instances are not wanting of roads which never require to be scraped, but they occur only where materials such as the pebbles of Warwickshire and Staffordshire and other similar hard substances are used, and where the road has a perfect exposure to the sun and wind.*

The road-men should scrape from the centre to the sides; the mud should not be scraped into or allowed to remain in the channels, as is too frequently the case, but should be put into small heaps, about one foot from the side channels, so as not to stop the running of water in them.

These heaps should always be removed the moment the mud is sufficiently dry to admit of its being put into carts or barrows.

The scrapings should never be laid in heaps on the wastes or footpaths, but should be spread

* For a description of a new scraping machine, see p. 259.

evenly over the hollow parts of the wastes, till they are brought to a regular surface, and afterwards they should be carted at once off the road to some convenient place till they can be otherwise disposed of. To do this effectually, when the materials are weak, large depôts should be provided on the sides of the roads, about four times the size of the depôts which have been proposed for holding materials.

Constant attention on the part of a road surveyor is necessary to keeping hedges clipped, and the branches of trees in the fences lopped. The hedges should be cut so as to be as low as they can be kept without making the fence unfit for confining cattle within them. The value of a full exposure to the sun and wind in contributing to the preservation of roads is shown by the superior condition at all times of roads crossing uninclosed land.

The trustees of a turnpike road should require their surveyor to lay before them, at the commencement of every year, an estimate of the work he proposes to execute in the ensuing year. In this estimate every particular should be specified; namely, the quantity of materials to be provided, the prices to be paid for them, the labour to be employed, &c., &c. The surveyor should be required to make up an account at the end of every month of the money received and paid by him; and he should also make up an annual account showing the particulars of the year's expenditure, the quantity of materials bought and carried to the road, the sums paid for day labour, for task work, and for cartage, &c.

In some cases the practice of employing a pay-clerk has been introduced to pay for all the road expenses, in order to relieve the surveyor from all trouble about pecuniary matters, and at the same time to remove as much as possible all temptation to swerve from his duty. This practice has been attended with the best effects, and cannot be too strongly recommended.

CHAPTER XI.

ROAD INSTRUMENTS AND TOOLS.

THE principal instruments employed in surveying and laying out roads are theodolites, spirit levels, and sextants.

Theodolites.

Theodolites in careful and experienced hands are the best instruments for laying out a road, and for taking horizontal angles and intersections. The rates of inclination can be determined at once by means of the vertical arch, without any measurement by the chain being required: they are decidedly the best instruments.

Theodolites are made of various sizes and prices; but those that are five inches in diameter, and cost about 17*l.*, are the most suitable for road purposes.

These instruments are divided on the limb into spaces of thirty minutes, and by means of a vernier, single minutes can be read off with great precision.

They are furnished with a good telescope and spirit level, besides two levels on the limb set at right angles to each other, and a magnetic needle or compass in the centre, which is of use in getting the magnetic bearing of any line in the survey, or of taking the bearings independent of the divisions on the limb.

The theodolite is used in the following manner in surveying a road: —

When the line of direction is fixed upon, the theodolite is set up over the first point in the survey; it is then adjusted by means of the spirit levels, so as to be perfectly level. The eye piece of the telescope is moved in or out until the hairs are seen distinctly; and the object glass is adjusted to distinct vision, according to the distance of the levelling staff from the instrument. Zero on the limb is then brought to coincide with zero on the vernier plate, and the limb and plate are then clamped together. After this is done, the whole head is turned round until the north point on the compass box coincides with the north point of the needle. The limb is then screwed fast, and the vernier plate unclamped and turned round until the staff is seen through the telescope: the vernier plate is then clamped, and the observation completed by turning the tangent screws of the limb and of the vertical arch, until the centre of the vane exactly corresponds with the centre of the cross hairs in the telescope. The degree and minute on the limb and vertical arches are then read off and entered in the field book.

The distance from the instrument to the staff is then measured by the chain, and all offsets are at the same time measured and entered in the book. The length of the distance line is then carefully entered, and the theodolite removed and again set up directly over the point previously occupied by the levelling staff: this may be done by means of the plumb line usually attached to the instrument.

The next operation is to adjust the instrument perfectly level, and to send the staff back to the point originally occupied by the theodolite. The vane having been previously adjusted to the exact height of the centre of the telescope, the head of the instrument is then turned round until the staff is seen in the field of view of the telescope: the head is then clamped, and the bisection made by means of the tangent screw; the vernier plate still remaining steadily clamped to the limb. The vertical arch is then examined, to see if the degree and minute correspond with those previously observed: if not, the first observation must be repeated. The vernier plate is then unclamped, and the telescope turned round towards the next line of direction until the staff appears in the field of view; when this is effected, the vernier plate is clamped, and the observation completed as before. In this way the survey is carried on; and the perpendiculars and rates of inclination are afterwards calculated, and the plan and section laid down in the usual way.

Spirit Levels.

Troughton's levels, which are considered the best, are usually made with very powerful telescopes and delicate ground spirit levels. These instruments are usually fourteen inches long, but some are eighteen and others twenty inches long. They cost from 12*l*. to 18*l*., and are so well balanced and secured that they will not require, with proper care, for a long time, any adjustment.

The method of using these instruments is as follows: —

When the direction of the road has been marked out, a line is measured by a chain commencing at the beginning of the new line, and terminating at that point where the inclination of the surface of the ground changes, or where the line of direction changes. This distance is carefully entered in the field book. The spirit level is set up as nearly as possible in the middle of this line, and a levelling staff with a vane is held by assistants at each extremity of the line: the telescope is then adjusted for distinct vision, and its axis brought to be truly vertical, by means of the spirit level and parallel plates. The telescope is then directed to the staff, which is placed at the commencement of the line, and the assistant is directed to lower or raise the vane until it is bisected by the cross hairs in the telescope: the height marked by the vane on the staff is then set down in the field book in the column headed (back observation). The telescope is then turned round until the staff at the termination of the line is perceived in the field of view; the necessary signals are then given to lower or raise the vane on the staff until its centre coincides with the cross hairs in the telescope: the height of the vane on the staff is then entered in the field book in the column marked for observation, and the magnetic bearing of the line is also observed and set down in another column. Sometimes only one staff is used, in which case it is removed from the first to the second station after the observation is made. When very great accuracy is required, the level is set up by measurement exactly in the centre between the two staffs, for by this means the errors of adjustment and any slight deficiency in the instrument are compensated and mutually destroy each other.

Sextants.

The small pocket sextant is a most useful instrument in making road surveys; after a little practice, it can be used with great facility, and will be found a superior instrument to the common surveying needle, and much more accurate, besides affording the most expeditious method of making surveys of any yet known.

ROAD TOOLS.

Spades.

In some parts of the clay districts, a narrow spade, considerably curved in the blade, technically called a grafting tool (Plate VII. fig. 10.), is much used, particularly in cutting deep drains in stiff clay.

Shovels.

The best description of shovel for road work is pointed in the blade, and has a curved handle to allow the workmen to bring the blade flat to the ground without stooping. (See Plate VII. fig. 13.)

Trucks.

When metal rails can be laid down, the truck or small waggon is the best description of carriage for removing earth. A drawing of one of these is given in Plate VII. figs. 11 and 12: they usually hold a cubic yard of earth. The body is generally made of elm, the frame of oak, and the wheels and axles of iron.

Hammers.

Two descriptions of hammers, which are the most useful in road works, are represented in Plate VII. figs. 15 and 16. The handles should be flexible and made of straight grained ash, particularly those used for breaking pebbles: the small hammers should have a chisel face, and the larger ones a convex one, about five-eights of an inch in diameter. Those made of cast steel are the best; and though expensive in the first cost, they wear much better than wrought-iron ones, and very seldom break at the eye.

Pronged Shovels.

Pronged shovels are useful for filling stones, when broken, into carts or barrows: a drawing of one is given in Plate VII. fig. 7. A man is enabled to lift stones with much greater ease and more expeditiously with one of these shovels than with a common one; besides, he lifts them without taking up any earth with them.

Scrapers.

Scrapers are sometimes made of wood shod with iron, but those made of plate iron are preferable: they should be six inches deep, and from fourteen to eighteen inches long in the blade, according to the materials of which the road is composed; the softer and more fluid the mud, the longer the scrapers should be; they should be turned a little round at the ends, to prevent the mud from escaping. The best scrapers are made of old saw plates, stiffened on the back by a rib of wrought iron, or by riveting the plate to a board of elm, cut to the proper width and length, and about half an inch thick.

Patent Road Scraper.

The following is a description of a patent road scraper: —

"This machine consists of a series or row of scrapers placed in a frame, and mounted on wheels; it is worked crosswise on the road, and deposits the dirt or dust on the side.

"The machine is easily used by one man, and cleans above a mile of road per day, or about five times as much as can be done with the common scraper, and the work is better performed.

"The advantages from the use of this implement may be stated shortly as follows, and will be readily appreciated by surveyors: —

"1st. Improvement of the surface of the roads, which cannot be kept hard and firm unless the road be frequently scraped.

"2d. The facility afforded to fast travelling, by removing a great obstruction to the progress of carriages.

"3d. The preservation of the roads, by removing the dirt, which absorbs and retains water on the surface.

"4th. Assistance rendered to the surveyor, by enabling him to take advantage of favourable periods of weather for cleaning his roads.

"5th. The saving of money, which will be applied in strengthening or otherwise improving the roads.

"Wherever it has been tried, the machine has given great satisfaction, and the patentees are willing to send machines on trial." *

---

* Messrs. Bourne and Harris, of Ilchester, Somersetshire, are the patentees of this machine.

The following letter respecting it is from the engineer of the Holyhead Road: —

"Sir, Holyhead, 3d March 1837.

"On receipt of your letter this morning, I spoke to the foreman, and also to the labourer who attends to the road round this harbour, as to the comparative merits of the old and new scrapers: both of them gave the preference to the latter, as doing its work quicker and better than the former.

"When last I was at Corwen, Hugh Roberts the inspector told me that two of his men had had a dispute as to which of the scrapers was the best; they set to work, one with the common, the other with the patent scraper, and the result was, that the patent scraper took the dirt off cleaner than the other, as was proved by that part of the road on which it had been used remaining for a longer time without requiring to be scraped again than that part on which the common scraper had been employed.

"I am, Sir,

"Your most obedient servant,

"John Provis."

Hedging Knives.

These instruments have been long used in Scotland, where they are called plashing tools: they are made of different sizes; that represented in Plate VII. fig. 14. is the most useful. When a labourer is a little practised in the use of them, he can trim a hedge as well as a gardener with a pair of shears, and much more expeditiously. They should be made sufficiently light to enable a man to use them with one hand, and care should be taken by the maker that they are properly balanced on the handle, otherwise a workman will not be able to wield them with proper effect. The great error in making these instruments in England is that of making them too heavy, and curving the blade too much.

Working Levels.

Working levels are absolutely necessary in laying out new works, and in repairing old roads. These instruments are easily used by common workmen. One of the best kind of these levels is represented in Plate VII. fig. 8., in which A B C represents the level, upon the horizontal bar of which are placed four gauges, *a*, *b*, *c*, *d*, made to move perpendicularly to the line A C, in dove-tailed grooves cut in the horizontal bar. When any of these are adjusted to project a proper depth below the line A C, it may be fixed by a thumb screw, which will retain the gauge in the desired position.

Fig. 9. shows a section of the horizontal bar drawn to a larger scale, as marked upon the edge of the gauge. This section is taken through the line *e f* of fig. 8. In this figure the position of the square iron bolt, or screw pin, is more plainly seen, and also the washer placed under the thumb screw. Three of these bolts pass through the horizontal bar, fig. 8., exactly three inches above the line A C; the other, seen at *d*, is only two inches above the same line.

Levels for laying out slopes are best made of a bar of wood, three inches deep, one inch thick, and six feet long: on the centre near the middle of the rod, a triangular piece of wood of the same thickness is nailed; the sides of this triangular piece are so formed, that when the rod is placed upon a slope of one to two or one to three, a small pocket level placed on one side of the triangle will be horizontal, and the bubble will remain in the centre.

Ring Gauges.

Ring gauges for ascertaining the size of the broken stones are extremely useful. A ring of this description is represented in Plate VII. fig. 17.

CHAPTER XII.

ROAD LEGISLATION.

Turnpike System.

According as the trade and wealth of England increased, the roads became wholly unfit for the traffic on them, and this led to the introduction of turnpike tolls and boards of trustees for collecting the tolls and superintending the roads.

The legislature proceeded on a principle that was a perfectly just one, and, at the same time, in regard to its efficacy as a means to an end, one in every way judicious and right. For nothing can be more just than to make those who use the roads find the money for maintaining them, and no plan for obtaining such a large amount of money as is necessary for maintaining them in good condition could have been adopted more effective for that object. If rates on land had been resorted to, the measure would inevitably have failed, because the landowners would, beyond all doubt, have preferred bad roads and low rates to good ones and high rates: in point of fact, very indifferent roads would have answered all their local purposes. If the roads had been vested in the hands of government, it may safely be said, that this plan would also have failed, for government would never have been able to obtain the consent of Parliament to vote upwards of a million and a half a year for those roads only which now are turnpike roads. It is therefore to the turnpike system of management that England is indebted for her superiority over other countries with respect to roads.

The legislature, by providing for the levying of tolls, and giving powers to persons willing to come forward as subscribers, commissioners, or trustees, and act together for the purpose of making new or improving old roads, adopted the wisest principle for securing an abundance of them.

Had the legislature depended on the public treasury or on rates on property for funds, or had it refused to incorporate those persons who have executed the duties of turnpike trusts, and given the management of the roads to the government, or left them wholly with the parishes, this country could never have reached the degree of wealth and prosperity to which it has arrived, for want of proper means of inland communication.

It must be clear to every one who has carefully examined this subject, that nothing but leaving the management of the roads to those persons who live in their neighbourhood would ever have induced the people of England to pay, as they now do, a revenue, arising from turnpike tolls, to the amount of 1,400,000*l.* a year; for, although tolls are in every respect fair and proper for the purpose to which they are applied, and although government, by employing scientific engineers, might have expended the produce with greater skill than country gentlemen, the hostility to pay them, had they been wholly at the disposal of government, would no doubt have prevented their being generally introduced, and the making of useful roads over the whole country so universally as they have been made under the established system.

It should be remembered, that turnpike roads owe their origin, in many instances, to private subscriptions of considerable amount; and, in every such case, the main inducement to subscribe must have been that of entrusting to the subscribers the management of the funds, and giving them corporate powers.

The same principle of association has led to the construction of the canals, the docks, the great bridges, and all the most useful public works of the country; and it is not conceivable how such large funds as have been subscribed for making new lines, or for placing parish roads under turnpike trusts, could have been obtained as have been obtained, had not the legislature acted on this principle.

The following statement has been taken from "An abstract of the income and expenditure of the trusts of England and Wales, from 1st January 1835 to 31st December 1835, pursuant to the act 3 & 4 Will. IV. cap. 80." [Par. papers, session 1837, No. 328.]

	£
"Revenue received from tolls -	1,469,317
Money borrowed on security of tolls	165,474
Manual labour - - - - -	397,665
Team labour - - -	134,861

	£
Materials for surface repairs -	215,835
Interest of debt - - - -	301,508
Improvements - - -	211,808
Debt paid off - - - - -	132,983
Bonded mortgage debt -	7,116,702

Number of trusts, 1,111."

It would appear from this statement, that the management of the finances of the trusts was improving. Against the sum borrowed in 1835, of 165,474*l.*, is be set the sum expended in improvements, of 211,808*l.*, and the sum paid in reducing debt, of 132,983*l.*, leaving a surplus of 179,317*l.*, arising from care and economy. The great change for the better, within a few years, that has taken place in the manner in which trustees attend to their duties, has introduced an improved system of management, which will, no doubt, lead to the gradual reduction of the debt. In those cases where neither principal nor interest can be paid, there does not appear to be any just claim on the public at large, upon the part of the suffering creditors. Many of them have received compensation for their money by the improvement of their estates, and the others stand in the same position as all creditors do who lend money on speculation.

But although it is unquestionably true, that it is to the turnpike system that the abundance of useful roads is owing, it must at the same time be observed, that great errors have been committed in carrying the system into operation. For however numerous and however useful the roads may be, they are, as has been already stated more than once, extremely imperfect, in comparison with what they ought to be.

In respect to the lines of direction, it has been observed that they are every where extremely faulty. They have commonly been carried over all the hills between the points of communication, whereas they might have been kept on comparatively level ground along the valleys of the country.

While the most magnificent improvements have been going forward in all other kinds of public works, displaying the greatest efforts of human skill, and a rapid advancement in the science of civil engineering, scarcely any road can be pointed out, except a few which have been put under the management of civil engineers, that is not defective in the most essential particulars.

Who is to blame for this? Not the government, because the business has not been in its hands. The leading men of the commercial and manufacturing classes, who have been chiefly concerned in forming companies for making canals, docks, bridges, and other splendid improvements, are not to blame, for they have been too generally excluded from the business of road management. Nor are the civil engineers of Great Britain to blame, because they have seldom been consulted: on the contrary, this profession has been too commonly deemed, by turnpike trustees, as something rather to be avoided, than as useful and necessary to be called to their assistance.

The country gentlemen of England, in point of fact, are alone responsible for the defective state of the roads, because the business of managing them has been vested by the legislature exclusively in their hands.

Dr. Adam Smith bears testimony to the bad management of road trustees in his time. He says: "The money levied is more than double of what is necessary for executing, in the completest manner, the work, which is often executed in a very slovenly manner, and sometimes not executed at all." This remark, in too many cases, is just as applicable now as it was when first made, sixty years ago.

In those instances where a turnpike road is used merely for local purposes, however defective it may be, those persons only are put to inconvenience who live near it; but, where a turnpike road forms the communication between populous cities or towns at a considerable distance from each other, then the misconduct of trustees, whether arising from negligence, ignorance, or corruption, is of serious importance, and calls loudly for correction and control.

We shall now proceed to state what appear to be the principal errors in legislation which have been committed in giving effect to the turnpike system.

According to the provisions of every turnpike act, a great number of persons are named as trustees: the practice is to make almost every opulent farmer or tradesman a trustee, residing in the vicinity of a road, as well as all the nobility and persons of large landed property; so that a trust seldom consists of fewer than 100 persons, even if the length of the road to be maintained by them does not exceed a few miles. The result of this practice is, that in every set of trustees there are to be found many persons who do not possess a single qualification for the office; persons who conceive they are raised by the title of a road trustee to a station of some importance, and who too often seek to show it, by opposing their superiors in ability and integrity, when valuable improvements are under consideration; taking care, too frequently, to turn their authority to account, by so directing the expenditure as may best promote the interests of themselves or their connections.

It sometimes happens that if one trustee, more intelligent and more public spirited than the rest, attempts to take a lead, and proposes a measure in every way right and proper to be adopted, his ability to give advice is questioned, his presumption condemned, his motives suspected; and as every such measure will almost always have the effect of defeating some private object, it is commonly met either by direct rejection, or some indirect contrivance for getting rid of it. In this way intelligent and public spirited trustees become disgusted, and cease to attend meetings; for, besides frequently experiencing opposition and defeat at the hands of the least worthy of their associates, they are annoyed by the noise and language with which the discussions are carried on, and feel themselves placed in a situation in which they are exposed to insult and ill usage.

Numerous cases could be quoted to prove the accuracy of what is here stated; but it is unnecessary to do so, because every one acquainted with the subject who reads these remarks will readily allow their general correctness, and be prepared to admit that the sketch might easily have been still more highly coloured.

There is one effect of having these large bodies of managers, which is particularly deserving of notice, and that is, the necessary want of uniformity and system in their measures. It often happens that, when some important business is to be performed, one set of ten or twenty trustees, after devoting a great deal of their time in attending meetings, finally decide upon some useful measure, when another set of trustees summon a meeting, and rescind all their fellow trustees have done. This is a course of proceeding which is, of itself, sufficient to establish, beyond all dispute, the absolute necessity of some considerable change in the existing system.

Notwithstanding that the state of the turnpike roads was inquired into by select committees of the House of Commons in the sessions of 1819, 1820, and 1823, and that in consequence of their reports a new general turnpike act was passed in 1823, the evil of the mal-administration of the powers of trustees has not been cured by the 153 clauses contained in this act. The evil, in point of fact, having its source in the principle on which the body governing road business is formed, is not of a nature to be cured by a multitude of regulations; and the framers of the law committed a great error in overlooking this point. It is the principle of having such a number of trustees, all possessing the same powers, that throws every thing belonging to road operations into confusion, and produces the waste of the road funds. A law, therefore, to do any good, should provide that the number of trustees shall be limited within some rational bounds, or that the executive duties should be vested in a small body of them.

The committee of the House of Commons appointed in 1823 to inquire into the state of the turnpike roads say, in their report,—"Your Committee would therefore strongly recommend to the House the consideration of the subject of making and managing the roads of the kingdom in the course of the ensuing session of parliament; feeling convinced, that whatever plausible appearance the plan may assume of appointing a large number of noblemen, gentlemen, farmers, and tradesmen, commissioners of the roads, that the practice has every where been found to be at variance with the supposed efficiency of so large a number of irresponsible managers; and that the inevitable consequences of a continuance of this defective system will be to involve the different trusts deeper in debt, and leave the roads without funds to preserve them in proper order."—(*Report*, p. 9.)

But besides diminishing the number of trustees,

other steps should be taken in order to secure a uniform and efficient system of management of the executive business of maintaining a road.

The first measure that should be taken is to alter the present plan of having road business transacted at boards of trustees at which every trustee has a right to attend and vote. Experience proves that no contrivance can be worse for managing business than a board consisting of numerous commissioners or trustees possessing equal powers. This subject has been recently examined into by the commissioners of excise inquiry, and as what they say upon it, in their twentieth report, is in every respect applicable to turnpike trusts, the following extract has been taken from that report:—

"In submitting this topic for consideration, we should add, that the throwing aside of control is not the only inconvenience that arises from placing the administration of public business under a board; and that after having directed much attention to an examination of the question, as to the most eligible mode of conducting those portions of the public business which are at present intrusted to the management of boards, we have come to the conclusion that there are defects inherent in the nature of boards, constituted like those of the principal revenue departments, which prevent the affairs intrusted to them, and more especially those which relate to the discipline and control of large classes of subordinate officers, from being conducted with that degree of dispatch and efficiency, which are exerted when the governing powers are in fewer hands, and exercised under a less divided responsibility. In speaking of the boards thus constituted, as consisting of numerous commissioners exercising equal powers, we are aware that the numbers of each board have been of late years much reduced: still, however, a board composed of even six or seven members must be considered as a numerous body for the dispatch of executive business; and it may be added, that although the consequence of the reduction of numbers may have been the exertion of an increased degree of diligence and attention on the part of the individual members of whom the boards now consist, still that the defects in the general system of managing their business continue to be much the same in character, though perhaps less in degree, whether the boards consist of twelve members or of six. Amongst these defects, experience has shown the existence of a general disposition to avoid or postpone the discharge of those branches of the duty of a governing authority which require the exertion of individuals; and on the other hand, a disposition to make continual references to subordinate officers to obtain knowledge and information. The consequence of such habits are continual delays and adjournments of business; waste of time in debates; and, instead of that activity and energy, (which should be the characteristics of a body on whose superintendence depends the due administration of a complicated branch of laws and regulations requiring constant attention and variations in practice to adapt them to the objects for which they are to be enforced,) it has been found to be a prevailing custom to allow business to run its course according to routines of ancient regulation. Changes and reforms of any large and therefore useful kind are avoided; means, that have been provided at a great expense for the checking of the violation of the laws and the protection of the revenue, are not brought into action. The powers of boards almost always fall into the hands of some clever and active subordinate officer; while the silence and secrecy with which their proceedings are carried on leave the public uninformed of them, and thus allow abuses to grow up and go on to a great extent without correction. In confirmation of what is here stated, the preceding pages of this report, and the evidence annexed to it, may be referred to, as showing that, in the case of the board of excise, the first class of surveying-general-examiners directs many of its principal proceedings: the under secretary, from his long experience and abilities, has necessarily a preponderating influence; the routine of business is carried on with a strict adherence to old forms; and although expensive schemes for suppressing smuggling have been adopted, they have not always produced much effect, for want of activity and energy in enforcing them. Throughout the department there is much room for more uniformity, simplicity, and economy both in time and expense, whilst the general course of its proceedings does not appear to be governed by anything like a system of fixed principles, such as we might expect to find established for regulating the administration of a great public department.

"There is another great defect to be noticed belonging to the management of business by boards, and that is, the depriving of the public of the security of personal responsibility for the proper performance of its business. In the case of managing the excise duties, the First Lord of the Treasury and the Chancellor of the Exchequer are relieved in appearance from responsibility by a board appointed with full powers to collect them, and the public adopt what exists in appearance as a reality, and do not consider them responsible.

"The responsibility of the board, as a board, is of no value whatever; and, as to the commissioners individually, no one of them is responsible for the acts of the board, as others participate with him in all he does, and as much may be done in which some member of the board has not acted; so that in fact the appointing of a board of several commissioners with equal powers, as the head of a sub-department for revenue purposes, completely sets aside all responsibility.

"Experience of managing business by boards (a system which in this country is so common) affords a complete illustration of the correctness of the preceding observations. The proceedings of the numerous boards of commissioners for paving and lighting, and of sewers, are seldom mentioned but in terms of complaint and condemnation. The conduct of vestries, which are boards of a more extended kind, produced so much evil while they had the management of the poor that it led to their being set aside by the new Poor Law; and such has been the general bad management of commissioners of turnpike roads, that, by common consent, parliament is called upon to introduce some great change in the system. When a board is composed of numerous members, many of them have too many occupations, and many are too indolent, or of too much dignity, to attend to the business of it; and thus the apparent management by the whole body becomes a screen for the measures of a few, into whose hands the management practically falls. Thus it may happen, that for want of attendance, want of intelligence, want of economy, or want of some other requisite in the quarter to which the actual management has been left, the most lavish and wasteful expenditure of funds may take place, and the interests of the public sacrificed in this and a number of other ways.

" As to the fitness of a single person to discharge the duties of the head of a revenue department, so far as fitness consists of intelligence, activity, and energy, this would be insured in a higher degree in a single person than in a board. The exertion which an individual would be called upon and be able to make would secure a higher degree of intelligence, while the admitted superior force of individual activity and energy would not be weakened by waiting for the opinions of others, by debating with and gaining over others, by being driven to take unnecessary steps, and by adjournments of business."*

The commissioners have made out their case so completely, of the unfitness of boards to manage business, that it is unnecessary to say anything more on the subject. All that is known of the proceedings of turnpike trusts corroborates the accuracy and soundness of the reasoning and conclusions of the commissioners.

In order to remedy this evil of board management of roads, a law should be made containing the following provisions: —

1. That every trust shall elect, by ballot, a chairman, a deputy chairman, and three more to form a committee of trustees.

2. That in each year every surveyor of a road shall lay before the trustees of it an estimate, showing the work to be done in the ensuing year, and the expense to be incurred.

3. That the trustees shall immediately proceed to examine the estimate, and to declare what works shall be executed in the ensuing year, and what expense shall be incurred upon them.

4. That the chairman of the committee shall have the sole direction of the execution of the works, and shall be empowered to give orders for paying for them.

5. That the chairman shall have the power of appointing and dismissing surveyors and labourers.

* Twentieth Report of the Commissioners of Excise Inquiry, p. 128.

6. That the chairman shall have the power of letting tolls; but no letting to be final without its having been first approved of by a board of trustees.

7. The chairman, deputy chairman, and committee to remain in office for three years. In case of vacancies the trustees to meet and fill them up.

8. The deputy chairman to act for the chairman in case of sickness, or absence from other causes.

9. The committee to assemble whenever they are summoned by the chairman, to deliberate and give their advice upon whatever subjects he brings before them.

10. A salary to be paid to the chairman.

11. The chairman annually to lay a report of his proceedings before the trustees.

If this plan were adopted, the following practical advantages would follow:—

1. The expenditure would be founded on well prepared and well considered estimates.

2. Instead of the management and direction of the expenditure being, as it now is, practically in the hands of the clerks and surveyors, it would be vested in persons chosen for their qualifications to transact such business by the free election of the trustees.

3. The responsibility that would be thrown upon those persons would induce them to study the science and art of road-making, and no doubt, in a short time it would be seen, not only that the expenditure was more economically conducted, but that scientific management would supersede the existing system, under which the roads are so des-

titute of neatness and uniformity in appearance, and so defective in solidity and hardness.

4. The business of controlling the letting of the tolls, and of deciding upon the works to be executed, would remain in the body at large of trustees.

It is only by legislating on principle, and in this way founding an efficient governing authority, that any general and useful reform can be secured of the existing turnpike trust system. So long as legislation, overlooking the condition of the fitness of the governing authority to the end in view, seeks to introduce reforms by new regulations of control to be brought into execution by boards of trustees, it will fail, in consequence of its not having attacked the cause of evil, namely, the vicious formation of the governing authority.*

Another great evil of the existing system, which a new law should correct, is that of placing a line of road under the management of too many separate boards of trustees. With respect to cross-country roads, it may be difficult to apply a remedy to the evil; but as to all the mail-coach roads of the kingdom, a law should be passed to provide for

* Taking into consideration the whole of the circumstances of the turnpike roads, it would appear that no plan could be less adapted to the improvement of them than that contained in the bills which were brought into the House of Commons for consolidating the trusts in the last two years. They did not contain any regulations founded on the right principles of the science of administration, as applicable to the management of road money, and to the exercise of authority over officers and workmen.

the consolidation of existing trusts, by voluntary arrangements, when they can be made; or, by taking advantage of the expiration of the existing acts, to vest the roads to which they relate under the management of the adjoining trusts, so as to have at least fifty miles in each trust.*

* Extract from Mr. Telford's first annual report on the Holyhead Road, dated May 4th 1824, p. 25:—"Perfect management must be guided by rules and regulations; and these must be carried into effect by the unceasing attention of a judicious and faithful surveyor, who has, by actual experience and attention, acquired a thorough knowledge of all that is required and applicable to the general and local state of particular districts, as regards soil, materials, and climate, likewise the sort of wear to which the surface is liable. A person possessed of all these requisites, and otherwise properly qualified to level and set out new lines, &c. where necessary, must receive the remuneration such a character merits, and may always obtain, in this active and industrious country. But however convinced and well disposed trustees may be to give this remuneration, the tolls of five or six miles do not afford the means of giving it. The consequence is, that the Shifnal trust (four miles) has hitherto been under the management of a person so little acquainted with proper road business, that it becomes a serious consideration whether it will be prudent to suffer the extensive improvement at Priors Leigh to be entrusted to his care. Until the parliamentry commissioners interfered, and showed a practical example, the Wellington trust (seven miles) was managed almost wholly by the clerk: he had a sort of foreman, who appeared to be only partly employed on the road. And on the Shrewsbury trust (seven miles), as has already been stated, the surveyor and contractor were united in the same person. All these managers proceeded without regard to any rules and regulations whatever, receiving only occasional directions from some of the most active of the trustees, whose varying opinions served more to distract than benefit the prac-

Another very great defect in our legislation remains to be noticed, namely, the want of some power to control the trustees of turnpike roads, and to prevent neglect and corrupt practices. No other trustees are free to do whatever they please with perfect impunity; and no reason can be given for not making every one who takes upon himself such an office accountable before a proper tribunal for his conduct in the discharge of its duties. Dr. Adam Smith has remarked the great defect in the turnpike laws of not providing such a control. Mr. Burke says, "It is of the very essence of every trust to be rendered accountable, and even totally to cease when it substantially varies from the purposes for which it could have a lawful existence." If a board of trustees suffer the road under their care to get into a bad condition, the only remedy is to indict the parish through which the road passes; but nothing can be more contrary to every principle of justice than such a law. In all cases where trustees have the management of landed property applicable to the maintenance of buildings, bridges, and roads, proceedings may be taken against them in the court of Queen's Bench, if they

tical operations of the workmen. I must beg leave to add, that these observations are applicable to all trusts of similar extent, and are evidence of the propriety of establishing districts of a magnitude to justify a more perfect arrangement, and the employing of a properly qualified surveyor, whose sole occupation should be the road under his care, and who should also be enabled to keep constantly employed a set of workmen thorougly conversant with road operations, and working chiefly by contract."

abuse the trust reposed in them. In the case of roads, the circumstance of the funds for maintaining them being derived from tolls should make no difference, and the trustees should be equally liable with those who have the management of estates to be brought before this court. But this remedy would not be sufficiently easy and efficacious. A more direct and ready course of proceeding would be to allow complaints against trustees to be brought by petition before the judges at assizes. The judges should be empowered to try, with a jury, the allegations contained in the petitions; and in case of a verdict in favour of the petitioners, the judges should be enabled to set aside the trustees, and name commissioners to take charge of the road for as long a period as they might think advisable.

In order to afford further protection to the public against the misconduct of trustees, the House of Commons ought not to allow turnpike bills to be passed as a matter of course. A particular set of standing orders should be framed for the purpose of keeping such persons in check. No bill should be allowed to be read a first time in the House of Commons, for renewing an act, until after a select committee had been appointed to examine minutely into the state of the road, and into the accounts of it; and time should be allowed for petitions to be presented to the House against the bill, and for having the allegations contained in them fully examined.

But in addition to the measures now proposed, however well adapted they may be for putting the trustees under more control than they now are, another should be taken, further to secure an upright and efficient discharge of duty, namely, that of placing them under a central board of commissioners, with powers of inspection and superintendence, but not with such powers as would essentially interfere with local management.

Although in principle the general policy of England is right, of leaving local affairs to local management, experience of the working of it seems to show, that, whenever the interests of the public at large are mixed up with local affairs, some degree of control, on the part of the public, ought to be established. When local affairs are of such a nature that they are confined to a vicinage of limited extent, the only persons who suffer from mismanagement are those who are guilty of the mismanagement, and no one need pity them, or try to make them manage better: their own i terests will lead in time to the proper remedy. But when, under the plea of managing local affairs, things are done, or omitted to be done, by which the public is incommoded and injured, then the local authorities ought to be brought to account, and their misconduct exposed and corrected. Now, all main turnpike roads, though commonly considered as of a local character, involve to a great degree some most important public interests, because the greatest number of those who make use of them are strangers to the districts through which they pass; and therefore the public should be protected by some controlling authority possessed of powers, at least, to ascertain and make known the proceedings of the local trustees.

The Commissioners of Woods, Forests, and Land Revenue are well suited to act as a board for this purpose. They have recently been appointed to do the business heretofore done by the Board of Works, and also to execute the powers vested in the Commissioners of the Holyhead Roads.

If the principal roads in England, Scotland, and Wales were placed under these commissioners, they should have power given to them to cause annual inspections to be made by competent civil engineers, so as to obtain accurate information concerning the proceedings of every turnpike trust. Every trust should be obliged to furnish them with an annual account of its income, expenditure, and debt, and they should also have authority to inquire into the particulars of these accounts. An annual report should be made to parliament by the commissioners, containing a summary of the information derived from their inspections and inquiries.

This board, in addition to what is here required of it as a board of control, should be enabled to assist the trustees in making alterations and improvements. It should be authorised to have surveys made of all the mail-coach roads of Great Britain. These surveys should show the ground plan of each road, its vertical longitudinal section, and the alterations and improvements that may be made in it. A copy of the survey of each road should be given to the trustees who have the management of it, and the board should be enabled to make an arrangement with them for carrying the necessary alterations and improvements into execution.

In order that the board may be competent to make such an arrangement, powers should be given to it to issue exchequer bills similar to those possessed by the commissioners (under 57 Geo. 3. c. 54.) for issuing exchequer bills for public works. Loans should be made to the trustees, and they should be permitted to lay on additional tolls, to pay interest at the rate of three per cent., and to provide a sinking fund for repayment of at least three per cent. more.

But the money raised by these loans should not be paid over to the trustees; it should be held by the board, and expended by it in making, by its own officers, the intended alterations and improvements.

The board should have power to purchase land, procure materials, and do whatever is necessary for making new roads; but these, when finished, should be given over to the trustees.

This is the plan which has been acted upon by the parliamentary commissioners in making the improvements on that part of the Holyhead Road which lies between London and Shrewsbury.

When the parliamentary commissioners took in hand the improvement of this road in 1820, the road between London and Birmingham was one of the worst in England. The consequence was, that nearly all the travelling between London and Birmingham was by Oxford, though the longest way by eight miles; but now, the travelling has since been transferred from the Oxford to the Coventry line; so that the plan now proposed, with respect to the prospect of its success, has the sanction of experience.*

If this plan for assisting trustees in improving the roads were applied in the first instance only to the principal mail-coach roads, the expense to be incurred by the central board in making surveys and inspections would be of moderate amount. These might be made by resident civil engineers, acting under a chief engineer. The salary of each resident engineer need not exceed 300*l.* a year. Four assistant engineers in England, and one in Scotland, would be able to do all the business necessary for making surveys and reports, until the central board should have to execute new works. The resident civil engineer, under Mr. Telford, Mr. John Easton, who conducted for several years all the works on the road between London and Shrewsbury, received but 200*l.* a year. He made a survey of the whole line; prepared all the plans, estimates, specifications, and drawings for the

* Mr. Huskisson, as chairman of the commissioners of Land Revenue, was, *ex officio*, chairman of the commissioners of the Holyhead Road. When the author proposed to him the plan of placing the trustees of this road under their control, he fully approved of it, saying that, if the plan succeeded, all the roads of the kingdom ought to be placed under a similar control.

improvements; inspected the contractor's work; and instructed the surveyors of the local trusts in carrying on the repairs of the road.

The following extract from the report of the Committee of the House of Commons in 1819, on the public highways, contains remarks which concur fully in principle with the recommendations now given for the improvement of the turnpike roads :—

"The importance of land-carriage to the prosperity of a country need not be dwelt upon. Next to the general influence of the seasons, upon which the regular supply of our wants, and a great proportion of our comforts, so much depend, there is, perhaps, no circumstance more interesting to men in a civilised state, than the perfection of the means of interior communication. It is a matter, therefore, to be wondered at, that so great a source of national improvement has hitherto been so much neglected. Instead of the roads of the kingdom being made a great national concern, a number of local trusts are created, under the authority of which large sums of money are collected from the public, and expended without adequate responsibility or control. Hence arises a number of abuses, for which no remedy is provided; and the resources of the country, instead of being devoted to useful purposes, are too often improvidently wasted.

"Your Committee do not mean, by these observations, to recommend that the turnpike roads of the kingdom should be taken into the hands of government, as such a measure is liable to various objections; more especially as it would be difficult to compel either the government or its agents to keep the roads in a proper state of repair; and as, in process of time, the roads might be considered rather as a source of revenue, than an accommodation to the public. But your Committee are perfectly convinced, that leaving matters in their present state is in the highest degree impolitic. They are of opinion, that a parliamentary commission ought to be appointed, to whom every trust should be obliged annually to transmit a statement of its accounts, to be audited and checked. Before these commissioners any complaints of improper expenditure, by which so many innocent creditors suffer, ought to be brought and inquired properly into. An annual report of the state of the turnpike roads of the kingdom ought also to be laid, by such commissioners, before his Majesty and both Houses of Parliament. Such a commission would not be attended with any expense to the public treasury, as a small poundage on the money received by the different trusts would defray all the expenses it could possibly occasion.

"Nor is this all the advantage that would be derived from the proposed establishment. Under the direction of such an institution, the necessary experiments might be tried, for ascertaining the best mode of forming roads, and the best means of keeping them in repair; the proper construction of carriages and wheels; and the system of legislative provisions the best calculated for the preservation and improvement of roads. All these are points which cannot be brought to the state of perfection of which they are capable, without some attention on the part of the legislature; nor by committees of the House, occasionally appointed, however zealous in the cause. Such great objects, which would add millions to the national income, and would increase the comfort of every individual in the kingdom, can only be successfully carried through by a great and permanent institution, whose whole attention shall be directed to that particular object; and who would take a just pride in accomplishing some of the greatest benefits that could be conferred on their country."

The expense which must unavoidably be incurred in making roads as roads ought to be made, is in many cases so great, that it is not possible to acquire sufficient funds by any rate of toll which would be submitted to; and therefore it becomes necessary to provide some plan for obtaining them by other means. When the improvement required is of a principal mail-coach road, the public is so much interested, that the counties should be enabled to levy rates, to be given in aid of the road tolls.* The mail coaches also should pay

* By the 45 Geo. 3. c. 43. power was given to the Treasury to advance to the grand juries of Ireland loans for making and improving mail-coach roads, to be repaid in instalments by county rates. Several excellent roads were made in this way, according to surveys furnished by the Post Office, and all the loans have been repaid.

tolls, not to the trustees, but to the central board, to be applied by it in making improvements. It might moreover be proper to apply a part of the revenue to this purpose, derived from the duties on post-horses and stage coaches.

All past legislation may be said to have failed in producing perfect roads, in consequence of most erroneous notions as to the cost of making a good road. The want of correct opinions with respect to what constitutes a good road has commonly caused the necessity of providing adequate funds to be overlooked. With the greatest economy and skill, it is seldom possible to make a long line of new road in a proper manner, with no other funds than the money raised on the credit of tolls.

PARISH ROADS.

The roads commonly called parish roads, in England, are generally in a very imperfect condition. This is owing chiefly to the law by which their management is placed under the governing authority of the vestries of the parishes through which they pass. Blackstone says, "In England every parish is bound, of common right, to keep the roads that go through it in good and sufficient repair, unless, by tenure of lands or otherwise, the care is conveyed to some particular persons."

The principle here established, of placing the highways of the kingdom under as many separate governing authorities as there are parishes, is, in every respect, repugnant to any thing like a sound principle of management; and, until it is abandoned, no efforts of legislation can prove successful in introducing any real improvement.

So long as this radical error in principle shall be recognised by parliament, to pass acts of parliament containing a multitude of new regulations will be labour in vain. The influence of the original cause of the evils which prevail will render them, as they have rendered hundreds of similar regulations, wholly abortive.

Legislation on the highways of England, to be productive of any practical good, must be founded on a more enlarged view of the subject; and it seems advisable that, instead of the governing authority of a parish, that of a county should be substituted, or, when counties are very large, that of a division of a county.

The reasons which may be given to support this general proposition are so obvious, that it is unnecessary to state them all in detail; a few only will be noticed. The first is, that the private interests of a vestry lead it to be satisfied with very imperfect highways. A road that will allow a waggon to be drawn upon it without much difficulty will answer the purpose of those who commonly compose a vestry. But such a road need not have any other qualities than two ruts for the waggon wheels, and a track-way for the horses. The second reason is, that the limited extent and funds of a parish will not admit of giving such a salary to a surveyor as will secure the services of a person educated in the principles of road management, and otherwise qualified for the office of surveyor.

Another great error in the system of parish management consists in the regulation by which a surveyor is appointed to act only for one year. This practice is founded on the vulgar notion, that for the management of roads no education is required; that it is not an art to be directed by skill and science. It is a practice which may be set down as having its origin in very rude times, and made familiar by long usage; but it certainly is one which ought to be abolished in the present enlightened state of society.

To legislate, therefore, on sound principles, the old custom of seeking to amend what is wrong, by laws containing a multitude of new regulations, must be abandoned: the country gentlemen who, as members of parliament, undertake the task of legislating on the subject, must look more to general principles; and, in order to succeed, they should no longer allow parish vestries to be the governing authority, or surveyors to be appointed to act only for one year.

The principal alterations introduced into the law affecting highways by the last highway act are as follows:—

The surveyor under the old law was virtually selected by the magistrates; under the new law he is chosen by the inhabitants in vestry assembled, and in places where there is no vestry, at a public meeting of the rate payers. The time of election is altered from the autumn to the spring, when the other parochial officers are chosen.

Statute labour and all compositions in lieu thereof are abolished; and the expense attending the repair of all highways is to be defrayed by a rate levied in the same manner as the poor rate.

Provision is made for the appointment of a board of managers in populous districts, to discharge the duties of surveyors.

Parishes may unite for the purpose of forming unions or districts for the management of the roads, and district surveyors may be appointed with salaries.

The law with reference to widening highways remains unaltered.

The mode of diverting and stopping up highways is entirely different from that under the old law, where two magistrates could in the first instance make the order, subject to an appeal to the bench of magistrates at Quarter Sessions. Now, after certain notices have been given, and certain prescribed forms observed, which must be certified by two magistrates, the order is taken to the Quarter Sessions, to be made in the first instance; and in cases of appeal the magistrates present are directed to empannel a jury to try the merits of the case.

Presentment of roads by magistrates and constables is abolished. The power of indictment is reserved for extreme cases, and a more summary process before two magistrates prescribed in ordinary cases where roads are out of repair.

The remaining provisions of the act, which refer to what may be called the police of the road, are assimilated as much as possible to the provisions of the general turnpike act.

CHAPTER XIII.

CARRIAGES.

As the expense of maintaining a road in good order depends in some degree upon the sort of carriages which are made use of, a few remarks on different kinds of carriages may be introduced with propriety into this work.

When a road has been made with very hard materials, and it has a very smooth surface, a wheel in rolling over it, even bearing a great weight, does very little injury; but when it has been made with weak materials, a wheel cuts and injures the surface in proportion to the weight it carries. The general ignorance of road-making on right principles led all those who first undertook to improve the roads to attempt to make bad materials do as well as good ones, by regulating the breadth of the tires of wheels, and by limiting the weight to be carried on them. The consequence was, a great quantity of absurd legislation, and no improvement in the roads; and this, simply because it was impossible to make a good road with bad materials. If roads were made sufficiently strong and hard, of a proper form, and kept well drained and scraped, the only case in which the legislature need interfere, is to prevent the injury that wheels do which have nails projecting above the tires of them. With such roads, the interests of carriers of all kinds would lead them to make use of no other than one horse carts, as in Scotland and Ireland, and then the loads would never be so great as to produce any injury to a perfectly good road.

The experience of the use of one horse carts in Scotland and Ireland shows, that a much greater weight can be drawn by horses when working single than when they are joined together. The reason of this is, that it is impossible to make two or more horses exert their strength so that each horse shall regularly and steadily draw its proper share of a load. The common load of a one horse cart in Scotland and Ireland is 30 cwt., exclusive of the cart, while the average load that a horse draws in an English waggon is no more than 15 cwt., exclusive of the waggon.

The most simple and effectual way of getting rid of the injury which very heavily laden waggons do to roads would be, to increase the rate of toll on any additional horse; for instance, if the toll was 4*d.* on one horse, it should be 10*d.* on two, 17*d.* on three, and so on.

With respect to carriages used for carrying passengers, it would appear that the business of building carriages not having been interfered with by the legislature has been carried on very much to the advantage of the public. In other countries, France for instance, it is otherwise, for the shape and slow rate of moving of a French diligence is wholly owing to a most absurd regulation respecting the breadth of the tires of the wheels. But although stage coaches are so built in England as to allow of travelling with safety and convenience, and at a cheap rate, the comfort of travellers might be increased, and the labour of horses diminished, if the body of a coach were made larger, the fore wheels made higher, the springs made longer and slighter, and the weight made to rest chiefly on the hind wheels. The following observations, which were drawn up for the Commissioners of Inquiry into the Post Office, will explain fully what is proper to be done on each of these points.

ON THE CONSTRUCTION OF COACHES.

In order to form right opinions on plans of coaches, we should begin by acquiring an accurate knowledge of the use of wheels. Nothing is more common than to meet with persons who have formed the most decided opinions on the construction of carriages, without having examined the properties of wheels, and who do not know that they are treated by mathematicians as mechanical powers, and that their properties, as such, are exactly settled, and admit of no dispute among men of science.

The following extracts have been selected from Hutton's Mathematical Dictionary, Mr. Davies Gilbert's Treatise on Wheels and Springs of Carriages, Rees's Cyclopædia, and Ferguson's Lectures on Mechanics, as sufficient to give as much information as is necessary for deciding what should be the height of the wheels of a coach, and whether the loading of a coach should be placed, for the most part, over the fore or over the hind wheels.

Extract from Hutton's *Mathematical Dictionary. Article "Wheel."*

"The height of the wheel is of material consideration, as the spokes act as levers; the greater the length, therefore, of the spokes, or, what is the same thing, the greater the height of the wheel, the more the labour of horses is diminished in drawing a carriage on wheels.

"Large wheels are found more advantageous for rolling than small ones, both with regard to their powers as a larger lever, to the degree of friction, and to the advantage in getting over holes, ruts, and stones, &c. If we consider wheels with regard to the friction on their axles, it is evident that small wheels, by turning oftener round, and swifter about their axles than large ones, must have much more friction. Again, if we consider wheels as they sink into holes, or soft earth, the large wheels, by sinking less, must be much easier drawn out of holes, and over soft earth, as well as more easily over stones and other obstacles, from the great length of lever, or spokes. This has been brought to a mathematical calculation by Desaguiliers, in his Experimental Philosophy, vol. i., p. 171.

"Hence it appears, then, that wheels are the more advantageous as they are larger, provided they are not more than about five feet in height; for when they exceed that height, they become too heavy in order to make them sufficiently strong.

"It would be much more advantageous to make the four wheels of a carriage large, and nearly of the same height, than to make the fore wheels much lower than the hind wheels, as is usually the case.

"It is accounted a great disadvantage in small wheels, that as the axle is below the bow of the horses' breasts, the horses not only have the loaded carriage to draw along, but also part of the weight to bear, which tires them soon, and makes them grow stiffer in their hams than they would if they drew more on a level with the fore axle.

"If the wheels were always to roll on smooth or level ground, it would be best to make the spokes perpendicular to the naves or to the axles, because they would then bear the load perpendicularly. But because the ground is commonly uneven, one wheel often falls into a cavity, or rut, when the opposite wheel does not, and then it sustains more of the weight than the other does; in which case it is best for a wheel to be dished, because the spokes become perpendicular in the cavity, and therefore have the greatest strength when the obliquity of the road throws most of the weight on them, while those on the higher ground have less of the weight to bear, and therefore need not be at their full strength.

"The axles of the wheels should be quite straight, and perpendicular to the shafts, or to the pole. When the axles are straight, the rims of the wheels will be parallel to each other, in which case they will move the easiest, because they will be at liberty to proceed straight forward; but in the usual way of practice, the ends of the axles are bent downwards, which always keeps the sides of the wheels that are next the ground nearer to each other than the upper sides are; and this not only makes the wheels drag sideways, as they go along, and gives the load a much greater power of crushing them than when they are parallel to each other, but also endangers the overturning of the carriage, when a wheel falls into a hole or rut, or when a carriage goes on a road that has one side lower than the other.

"Experience shows that narrow-wheeled carriages are more easily drawn than broad-wheeled ones."

Extract from a Treatise on Wheels and Springs for Carriages, by Davies Gilbert, Esq., M.P., F.R.S., &c.

"Taking wheels completely in the abstract, they must be considered as answering two different purposes.

"*First.* They transfer the friction which would take place between a sliding body and the comparatively rough uneven surface over which it slides, to the smooth oiled peripheries of the axle and box, whence the absolute quantity of the friction, as opposing resistance, is also diminished by leverage, in the proportion of the wheel to that of the axle.

" *Secondly.* They procure mechanical advantage for overcoming obstacles, in proportion to the square roots of their diameter, when the obstacles are relatively small, by increasing the time in that ratio, during which the wheel ascends, and they pass over small transverse ruts, or hollows, with an absolute advantage of not sinking, proportionate to their diameters, and with a mechanical one, as before, proportionate to the square roots of their diameters; consequently, wheels thus considered cannot be too large:—in practice, however, they are limited by weight, by expense, and by convenience."

Extract from REES's *Cyclopædia. Article " Wheel."*

" In concave wheels, the rims are uniformly made conical, which subjects them to many disadvantages. Every cone that is put in motion upon a plane surface will revolve round its vertex; and if force is employed to confine it to a straight line, the smaller parts of the cone will be dragged along the ground, and the friction greatly increased.

" When the fore wheels are much lower than the hind ones, many disadvantages attend this construction. A considerable force is lost that would be effectual if they were large. The carriage would go on much more easily if the fore wheels were as high as the hind ones, and the higher the better, because their motion would be so much slower on their axles, and, consequently, the friction proportionally diminished.

"There is an inconvenience which attends the usual method of loading carriages; for when a carriage is loaded equally on both axles, the fore axle must endure so much more friction. However, the carriers put the heaviest part of the load upon the fore axle, which not only makes the friction greatest where it ought to be least, but also presseth the fore wheels deeper into the ground than the hind wheels, although the fore wheels, being less than the hind ones, are with so much the greater difficulty drawn.

"By throwing the greatest part of the load on the small wheels the draught becomes doubly severe on the horses, who thus unnecessarily suffer for the ignorance and folly of man."

Extract from FERGUSON'S *Lectures on Mechanics.* Edition 1806.

"Large fore wheels are advantageous, both in horizontal and inclined planes.

"It must naturally occur to every person reflecting upon this subject, that the axletrees should be straight, and the wheels perfectly parallel.

"In some carriages which we have examined, where the wheels were only 4 feet 6 inches in diameter, the distance of the wheels at top was fully 6 feet, and their distance below only 4 feet 8 inches:

by this foolish practice the very advantages which may be derived from the concavity of the wheels are completely taken away.

"The direction of the traces should be inclined to the horizon when the horse is at rest, in order that it may be horizontal when he lowers his breast and exerts his utmost force."

Extract from a Pamphlet entitled "Cursory Remarks on Wheel Carriages," by MR. JOHN COOK, *of Liquorpond-street, Coach Builder.*

"With smaller wheels before than behind, the disadvantage is increased in proportion to the smallness of them; so that if small fore wheels must be used, the weight ought to be brought nearer to the hind ones."—p. 15.

"From the before-mentioned cause (the great weight over the fore wheels), the fore wheels are pressed so forcibly into the ground, particularly in bad weather, that they are in a continual hole, while the hind ones, capable of taking the greatest weight, have the least; which, added to the immense friction from dished wheels, together with the present favourite system of curtailing their height, and of oblique traction, nearly doubles the draught."—p. 24.

In applying the principles here laid down, it will be proper not to carry them further than experience will justify, and only to approximate, as much as possible, the construction of a coach to these principles; and therefore, as experience shows that a wheel of the height of 4 feet 8 inches affords very great assistance in drawing carriages, in consequence of the length of its spokes, or leverage, and can be made of great strength, without being very heavy, it is proposed in the plan now submitted for consideration that the hind wheels of a coach should be of this height.

The risk of overturning being greater or less, according to the height of a coach from the ground, it is proposed, in order to keep the body as low as possible, to fix the hind end of the perch below the hind axle.

As the running of a coach, whether well or ill, depends very much on the way in which the hind wheels follow the fore ones, and as this depends in some degree on the length of the perch, the perch should be as short as it can be made with safety, with regard to the greater tendency of a coach with a short perch to overturn than with a long one. On these considerations it is proposed that the length of the perch, between the centres of the fore and hind axles, should be 6 feet.

From what has been said about wheels, it is clear that the fore wheels should be as high as the hind ones; but as such a height would interfere with the turning of a coach, and as fore wheels as high as 46 inches are already, in some cases, in use, it is proposed that the fore wheels should be of this height. It is probable that the further trial of wheels of this height will show that still higher ones may be conveniently used.

As a space of 7 inches should be allowed for the

play of the springs, the bottom of the hind boot should be 7 inches above the axle.

Most stage-coaches at present have the axles 4 feet, or 4 feet 1 inch between the collars, and the arms are set in such a manner that the wheels are 4 inches wider asunder at top than at bottom. This drooping of the arms adds very much to the draught. The principal reason in support of this dish-form given to wheels is that it enables the wheels to bear sudden lateral thrust. If a wheel with spokes perpendicular to the nave fall into a rut, and if in that situation the nave be pushed outwards by the swing of the carriage, the chances are that it may be thrust through altogether, and the spokes be forced out both from the nave and from the fellies; whereas in a dished wheel the nave cannot yield outwards until the shock be great enough to burst the tire by the extension of the spokes. Another practical advantage of a moderate degree of dishing is, that it admits of the spokes yielding a little when the tire or ring is first put on, and retaining their tendency to spring back to their first position, by which the fellies are made to continue to bear against the tire, even after it has been lengthened by the blows it gets in running over hard pavement.

The fore and hind axles should be of the same length. The arms should taper one quarter of an inch, and be slightly bent, so as to bring an upright spoke. The fore and hind wheels should rest on the ground 4 feet 8 inches asunder from out to out. The dish given to the wheels should be so adjusted as to secure the following of the hind wheels in the track of the fore ones.

Another inconvenience which arises from a perfectly upright wheel is the throwing of the road mud into the coach-windows and on to the roof.

The thickness and length of the bearing part of the axle should also be attended to; in heavy stage-coaches the arms are made $2\frac{1}{4}$ inches in diameter, and 11 inches long. If the arms be made of the best case-hardened iron, and perfectly cylindrical, 2 inches in diameter is sufficient, according to mechanical principles, with a bearing of 7 inches. The total length of the arms should, however, be 10 inches, 3 inches of it being hollowed out for holding oil or grease.

The proper width of the tires of wheels must be determined by experiment. That there is some certain width of tire which will produce the least resistance to draught cannot be denied, but what that width should be is at present uncertain. There is one thing, however, very clear, namely, that, theoretically, the narrow wheel produces the least resistance on good roads; and this is so far borne out by practice, that the coach-masters have found it to be their interest to make the tires of all the fast coaches very narrow. Some of those coaches which travel very fast have the tires not more on the bearing than $1\frac{1}{2}$ inch, whilst the heavy or slow coaches have tires of $2\frac{1}{4}$ inches. The mails have tires of 1 inch and $\frac{3}{4}$. If a proper set of experiments were made, the breadth of tires would be ascertained for different states of the weather on different kinds of roads. As roads are at present constructed, narrow wheels are not so injurious as formerly; and, until certain data be obtained, it will perhaps be advisable to make the tires not less than $1\frac{3}{4}$, or more than $2\frac{1}{8}$ inches.*

The convenience of travellers has been very little attended to in arranging the size of the bodies, and the height and depth of the seats inside coaches.

The object should be to enable the traveller to sit in such a position as will admit of his performing a long journey with the least fatigue.

A slight degree of consideration will make it clear, that the seats should be so high as to give ample room for the legs, and thus allow the whole length of the thighs to press upon the seats, and be supported by them. If the legs have not room to be nearly upright, and are therefore thrown forward, the fore part of the thighs are elevated, and the weight of the body is made to rest on the upper part of them, so as to produce a partial strain on the muscles in supporting the body.

* In France, the law requires that every diligence, which, with its loading, weighs one ton and three quarters, shall have the tires of the breadth of $3\frac{1}{4}$ inches; and that every diligence which weighs three tons shall have the tires of the breadth of $4\frac{1}{4}$ inches. As the roads of France have of late been much improved, the very great inferiority of coach-travelling in that country to what it is in this may be mainly attributed to this very absurd law respecting the tires of wheels. If, as in this country, there was no law for regulating the wheels and weights of diligences, the roads of France would soon be covered with light fast travelling carriages.

In addition to this, the body is thrown backwards, and the weight of it is thereby supported with a considerable strain on the muscles of the back, which leads to fatiguing them. There can be no doubt that a person who sits on a high, deep, and broad seat, and in an erect posture, as he does naturally on a well-formed high chair, will suffer less fatigue than one who sits on a low seat, and in a reclining posture.*

For these reasons, the floor of the well should be 7 inches below the body of a coach; the seats should be 15 inches above the bottom of the well; the cushions of them should be 4 inches thick, the depth of them 18 inches, and the breadth of them

* This is too true of all sorts of stage-coaches on the English roads. Much requires to be done to improve the mail-coaches, so as to render them less fatiguing to travellers going long distances. At present, the whole weight of a traveller's body is supported on the projecting bones at the root of the spinal column, which itself is unsupported throughout its whole length; as, from the perpendicular back given to the seats, the shoulder-blades are the only parts of the trunk which can touch them. If in the new coaches the seats should be made of the breadth proposed, and if they be made 1½ inches higher in front than at the back, then the whole of the thigh will find support, and by the action of its muscles considerable relief will be given to the other parts. Much may also be done to give relief to the traveller, by improving the hand straps; long pendulous straps, with their loops adjustible to different heights (as in some of the French diligences), would afford great relief, by affording a rest for the elbow of an arm passed through them.—*Extract from a Letter of Sir John Robisson, Secretary to the Royal Society, Edinburgh.*

44 inches. The height of the roof from the cushions should be 3 feet 6 inches, and the width of the door 21 inches.

By making the depth of the seats 18 inches, and allowing 2 inches for the breadth of the back panel and inside stuffing, the breadth of the side panels, outside, will be 20 inches. They should slope from the seats to the roof with a rake of $1\frac{1}{2}$ inches.

According to the established law of mechanics, that the higher the wheel the easier a weight upon it is drawn, when a carriage has two sets of wheels, one of which is necessarily lower than the other, the body of such a carriage should be so constructed, and so placed over the wheels, that the greatest possible portion of the load should be over the hind wheels. When the carriage is loaded, the fore wheels should press but lightly on the ground, only in such a degree as is necessary to prevent the fore part of it from tilting up.

In order to place the weight as far behind as possible, the body should be set as far back as it can be set, with allowing for the opening of the door. By rounding the door at the bottom, it may be set 4 inches within the front part of the rim of the hind wheel.

The hind boot should extend to 4 inches outside of the hind wheel, and this will be found to make its length 36 inches.

The height of the hind boot should be 32 inches, and the breadth of it as great as it can be made, by setting the side springs as far as possible asunder.

If the side springs were fastened under the hind axle, the hind boot might be made of the whole breadth of the body of the coach.*

The bottom of the fore boot should be on the same level as the bottom of the hind one.

The height of it should be 32 inches.

To give sufficient space for the passengers to move behind the coachman's seat, there should be 12 inches between the body and the seat, and the seat of the coach-box should be 15 inches wide. The fore boot, at the bottom, should be 25 inches in length, from a line drawn from the widest part of the body, 30 inches in breadth, and 34 inches in height.

With a view of explaining how necessary it is to make use of long and pliant springs, the following extracts have been taken from the treatises of Mr. Davies Gilbert and Mr. Edgeworth.

Extract from the Treatise on Wheels and Springs for Carriages, by Davies Gilbert, Esq., M.P., F.R.S.

"Springs were in all likelihood applied at first to carriages with no other view than to accommodate

* One horse carts are in common use at Paris capable of holding 30 cwt., which are supported on horizontal springs, 5 feet long, fastened under the axle. A moderate-sized horse draws one of these carts with great ease when fully laden. There are also in Paris a number of omnibusses supported on springs in this way.

travellers. They have since been found to answer several important ends.

"They convert all percussion into mere increase of pressure; that is, the collision of two hard bodies is changed, by the interposition of one that is elastic, into a mere accession of weight. Thus the carriage is preserved from injury, and the materials of the road are not broken; and, in surmounting obstacles, instead of the whole carriage with its load being lifted over them, the springs allow the wheels to rise, while the weights suspended upon them are scarcely moved from their horizontal level. So that if the whole of the weight could be supported on the springs, and all the other parts supposed to be devoid of *inertia*, while the springs themselves were very long, and extremely flexible, this consequence would clearly follow, however much it may wear the appearance of a paradox, that such a carriage may be drawn over a road abounding in small obstacles without agitation, and without any material addition being made to the moving power or draught."

Extract from a Letter of DAVIES GILBERT, Esq.

"A carriage without springs, moving over a rough road, has to be lifted over obstacles, or out of depressions, and all the power expended in overcoming *inertia* is pure loss; but the force exerted in elevating the weight is in a great measure by the preceding or subsequent descent. Now, under the supposition in my paragraph, *inertia* would be destroyed, and it actually is so, by the springs now at present used, and by the smooth roads."

Extract from an Essay on the Construction of Roads and Carriages, by RICHARD LOVELL EDGEWORTH, Esq., F.R.S., M.R.I.A. *Second Edition.*

"So great is the advantage of springs, that they almost annihilate the resistance which that par of the load which rests on them would encounter without them, upon stony roads, or rough pavement.

"From the whole of these experiments, it appears that the advantage of springs increases with the increased velocity of carriages."—p. 118.

"Indeed the advantage of springs has, from common observation, been stated to be equal to one horse in four."—p. 126.

"Upon the whole, the application of springs to carriages, either for carrying burthens, or for pleasure, tends not only to the ease of the traveller—to the safety of goods that are carried—to the preservation of roads—and to the duration of carriages themselves,—but they also materially facilitate their draught."—p. 170.

As the throwing the body so far back as has been proposed will not admit of long side springs if four horizontal springs were used behind, it is proposed that there shall be only one cross spring, and that the fore ends of the side springs shall be fastened to the body of the coach. The fastening of them should be adjusted in such a manner as to make the whole of the weight of the body and hind boot, before they are loaded, press on the parts of the side springs which are behind the axle. By this plan of having only one cross spring the body of a coach will have a much steadier support than it has according to the present plan of two cross springs, and the risk of overturning will be diminished.

The side springs to be 4 feet 6 inches long, the cross spring to be 46 inches.

In order to keep the weight of luggage as far behind as possible, nothing but carpet-bags and light packages should be put into the front boot. As the top of the coach will be only 7 feet 6 inches from the ground, portmanteaus and other heavy luggage may be put upon the roof. If 3 feet 6 inches of the roof next the guard's seat be used for this purpose, nearly the whole of the weight of the luggage on the roof will be over the hind wheels. For still further securing the object of having the weight as far behind as possible, a strong iron should project one foot from the bottom of the hind boot, having an iron frame, 18 inches in length, fastened to it with a hinge, so that extra mail-bags and heavy luggage may be packed to the full height, if necessary, of the top of the guard's seat.

The splinter-bar should be set 3 feet above the ground: this will make the angle at the point of the shoulder of the horse, between the trace and an horizontal line, an angle of about 15 degrees.

Summary of the foregoing Plan for building a Coach.—(*See Plates* VIII. and IX.)

1st. The hind wheels to be 4 feet 8 inches high.

2d. The fore wheels to be 3 feet 10 inches high.

3d. The perch to be 6 feet long.

4th. The perch to be fixed under the hind axle.

5th. The axles to be straight from end to end, and the bottom of the hind boot 7 inches above the hind axle.

6th. The rim of the hind wheel to project 4 inches beyond the line of the opening of the coach door.

7th. The breadth of the side panels, between the door and the hind boot, to be 20 inches.

8th. The end of the hind boot to project 4 inches beyond the rim of the hind wheel. The height of it to be 36 inches, and the breadth 42 inches.

9th. The breadth of the door to be 21 inches.

10th. The breadth of the side panel, between the door and the fore boot, to be 20 inches.

11th. The length of the bottom of the fore boot to be 25 inches; the height of it 34 inches; and the breadth of it 36 inches.

12th. The side springs before to be 30 inches, and the cross springs 46 inches.

13th. The side springs behind to be 54 inches long, and the cross spring to be 46 inches. The fore ends of the side springs to be fastened under the body.

14th. The well of the body to be 7 inches below the body, and 29 inches from the ground; the seats 15 inches above the floor of the well; the cushions 4 inches thick. The distance from the cushions to the roof to be 3 feet 6 inches.

15th. The depth of the seats to be 18 inches, and the breadth of them to be 44 inches.

16th. The guard's seat to be 15 inches above the hind boot.

17th. The splinter-bar to be 3 feet from the ground.

18th. The hind irons to be 2 feet 6 inches long.

The advantages of a coach built on the plan proposed, over a coach built on the plan now commonly in use, will be as follow: —

1st. By having the fore wheels higher, the weight set as much as possible over the hind wheels, the perch shorter, and the springs longer, the labour of the horses will be very much diminished.

2dly. By making the body wider, deeper, and higher, it will be much more convenient for travellers; they will also have great benefit from the longer springs.

3dly. By fastening the perch under the axle, the body of the coach will be at a less height from the ground than is now commonly the case.

Comparative Size of a Coach built on the proposed Plan, and of a common Coach.

	Proposed Coach. Inches.	Common Coach. Inches.
Fore boot - - -	25	25
Panel - - - -	20	17
Door - - - -	21	21
Panel - - - -	20	17
Hind boot - - -	36	33
Length of body and boots	122	113
Height of body and well	61	56
Height of seats and cushions inside -	19	17
Depth of ditto - -	18	15
Height of roof over the cushions - -	42	39
Height of fore wheel	46	42
Height of hind wheel	56	56

From this table it appears that great improvements may be made without departing very much from the present plan of building coaches.

As the practice of dishing wheels, and bending the arms of the axles downwards, originated in some degree in the bad state of the roads, now that they have become so much improved the degree of dishing heretofore given to wheels should be reduced, and made wholly to depend upon experience, with reference to the wearing out of wheels, and the expense of renewing them.

Attention should be paid to diminishing, as much as possible, the size and weight of all wood and iron work in coaches. As all coaches are provided by contract, and as it is the interest of the contractors to make them as strong as possible, it is much more than probable that they are built much stronger, and consequently much heavier, than necessary.

Notwithstanding the roads are now no longer cut into deep ruts and holes, the coaches are all nearly as heavy as they were fifteen years ago. The Post-office, and all coach proprietors, should take care, in making their contracts with the builders, to have the weight of each coach carefully determined upon, and properly specified in the contracts.

Although science, if properly applied, will certainly lead to improvement in the construction of coaches, a series of experiments should be made, in order to determine, with complete accuracy, what is the full effect of wheels, springs, and good roads in diminishing the labour of horses.

Mr. M'Neill's invention of an instrument for trying the draught of carriages, which has been found to be perfectly fit for the purpose, now admits of such experiments being made with a certainty of leading to accurate results; and it is very important that they should be made. In point of fact, although the extracts which have been taken from works of science are quite sufficient to convince all persons who have received a scientific education that the fore wheels of a coach ought to be high, and that the greater part of the load should be placed over the hind ones, as it happens that few of those persons who are concerned in the directing of the building and in the building of coaches have ever applied themselves to scientific inquiries, so as to know either why spokes are called levers, and what the property of the lever is, or what the effects are of the friction of wheels in turning on their axles and in moving on roads, it is quite necessary that experiments should be made, so that, by showing how much work horses actually do in drawing different kinds of carriages, nothing shall be left wanting to expose the prevailing errors with respect to wheels, and the proper manner of loading coaches.

As Mr. M'Neill's instrument can be fixed to a coach, with the horses to it, what it shows is, the actual force or labour which they exert in drawing; and, therefore, the experiments made with this instrument are not liable to errors, like other experiments, where it is necessary to use a substitute for the real power.

The expense of making a proper set of experiments would amount to some hundred pounds: as, however, there exists nothing to make it worth the while of any private person to incur it, these experiments should be ordered and paid for by government. This small expenditure would soon be repaid by the saving which would be effected by diminishing the labour of horses in drawing stage-coaches, and, consequently, the expense, which now falls indirectly on the public in providing a sufficient number of them, and maintaining them.

The plan here proposed for a mail coach may be easily adapted to stage-coaches.

For a light post-coach the hind side springs should be fastened under the axle, and the hind boot made 42 inches wide; a second set of irons should be put above the door of the hind boot, and a gammon-board should be attached to the body.

For a heavy coach the perch should be made 6 inches longer; the fore boot also 6 inches longer; the fore wheel 4 feet high; and the hind wheel 4 feet 10 inches high.

In packing coaches the hind boot should be first packed, and then the hind irons. The fore boot should be used only for parcels and luggage to be dropped or taken up on the road.

APPENDIX, No. I.

DESCRIPTION OF THE ROAD INDICATOR, AN INSTRUMENT INVENTED BY MR. MACNEILL FOR THE PURPOSE OF ASCERTAINING THE DRAUGHT OF CARRIAGES, AND THE COMPARATIVE MERIT OF ROADS; WITH TABLES OF EXPERIMENTS, ETC.*

THIS instrument, which is described in the following pages, is capable of being applied to several very important purposes in road engineering, amongst which are the following: —

First, It affords the means of ascertaining the exact power required to draw a carriage over any line of road.

Secondly, It can be applied to compare one line of road with another, so as to determine which of them is the best, and the exact amount of the difference, as regards horse power, both for slow and fast coaches.

Thirdly, The comparative value of different road surfaces may be determined with great exactness.

Fourthly, It affords the means of keeping a registry, in a most accurate manner, from year to year, of the state of a road, showing its improvement or deterioration, and the exact parts in which such improvement or deterioration have taken place.

* This paper has been furnished by Mr. Macneill.

Y

PRACTICAL EXAMPLES EXPLANATORY OF THE FOREGOING STATEMENT.

1st, Let it be required to determine the expense of working a four-horse coach over the line of road from —— to ——, at a velocity of ten miles an hour. Suppose the instrument has been run over the road, and that it has been found that the average power required to draw a four-horse coach over the whole line amounts to 350lbs., and the distance equal to twelve miles. Let the average power which a horse should exert for eight miles a day, with a velocity of ten miles per hour, be assumed equal to 60 lbs., then $60 \times 8 = 480$ lbs., raised one mile in the day; and taking the daily expense of a horse equal to six shillings, we have 480 lbs. : 6*s.* :: 1 lb. : ˙15, the expense of horse power, exerting a force of one pound over one mile. Thence, $350 \times {}^{\cdot}15 \times 12$ miles $= 630$ pence, or 2*l.* 12*s.* 6*d.*, the expense of horse power required to work a four-horse coach per day over such a road.

2dly, Suppose it be required to determine whether it is more expensive to work a coach over the stage from A to B, or over the stage from C to D, both stages being exactly ten miles, and horse keep the same in both districts. Let the instrument be run over both stages, and suppose the average power thus determined to be 280 lbs. on the stage from A to B, and 320 lbs. on the stage from C to D, the difference is $320 - 280 = 40$ lbs.; and this difference will amount to $40 \times 10 \times {}^{\cdot}15 = 5$ shillings in horse power, in favour of the stage from A to B.

Again: suppose the stage from A to B, which is ten miles in length, to be compared with the stage from E to F, which is only eight miles in length, but more hilly, or having a worse surface. Let the instrument be run over each stage as before, and suppose the average power from E to F to be found equal to 500 lbs., whilst the average power over the stage from A to B is only 320 lbs., as this stage is ten miles in length, the expense of working over it will be $320 \times 10 \times \cdot 15 = 576$ pence; and the expense over the stage from E to F will be $500 \times 8 \times \cdot 15 = 600$ pence; from which it will be seen that less expense will be required to draw the carriage from A to B than from E to F, although the distance from E to F is two miles shorter than from A to B; and that the difference of expense will be $600 - 576 = 24$ pence, or two shillings per day for a four-horse coach.

3dly, Suppose it be required to determine the best surface on different parts of a road, which has been constructed on different principles or repaired with different descriptions of road materials. Let the instrument be run over each portion of the road, and the average power noted — also the rates of inclination, as shown by the instrument, or a spirit level — then reduce the average draught over each rate of acclivity to what it would be if it was horizontal; the comparison of the corrected draughts will show the friction arising from the surface in each case. Thus, suppose the average draught over a portion of the road, which has been repaired with gravel, and which rises 1 in 20, to be 250 lbs. The correction for 1 in 20 is 39·2 lbs. The friction of the surface and axles is therefore 250—39·2, or 210·8 lbs. (See 7th Report of Parliamentary Commissioners of the Holyhead and Liverpool Roads, published by order of the House of Commons, January, 1830.)

In the same way, suppose the draught over another portion of the road which rises 1 in 10, but which has been repaired with granite, is found to be 260 lbs. The correction for 1 in 10 is 78·4 lbs., therefore the friction of the surface, or what it would be if it was horizontal, would be 260 — 78·4, or 181·6 lbs. only; the difference between this and the gravel surface will therefore be 210·8 — 181·6, or 29·2 lbs., which is equal to a saving of 4½ pence for every horse drawing over a mile of such a road, as compared with the other.

4thly, The most important and useful application of the instrument is, perhaps, that of being able to ascertain with accuracy and precision the state of any road, from time to time, as regards its surface; and the state of repair in which it has been kept.

The following table, or yearly register of a quarter of a mile of road, will show this more clearly. The numbers in the column represent the draught or horse power, taken at every ten yards. Thus, in the first column of the year 1829, the draughts were in summer 20, 30, 25, &c., and in the second, or winter column of the same year, the corresponding draughts on the same identical part of the road are found to be 35, 35, 30, &c.: these columns added up, and divided by the number of observations, give 44·5 lbs. for the mean summer draught, and 49·45 lbs. for the mean winter draughts, over this quarter of a mile. By following the same process in the following year, viz. in 1830, the mean summer draught was found to be 35·6 lbs., and the mean winter draught 40·36 lbs., showing that the road had been improved in the course of the year very considerably; and by a reference to the numbers in the columns on the same horizontal lines with each other, it will be found the improvement has been general, throughout the whole distance. In the next year, 1831, it will be seen that the average power in summer is 40·52 lbs., and in winter 46·5 lbs., which shows the road is not so good as it was in the preceding year, 1830, but better than it was in the first year, 1829. Again, in the year 1832, it is found that the average summer draught is 53·6 lbs., and the winter draught 63·18 lbs.: by comparing these numbers with any of the preceding years, it will at once be evident that the road has become worse; and by a reference to the figures in the column, it will be seen that it is defective in every part as compared with the preceding years, but more especially so near the end, where the draught in summer varies from 60 to 85 lbs., and in winter from 75 to 95 lbs.; whereas, in 1830, two years before, the draughts in summer, over the same part of the road, varied from 35 to 38 lbs. only, and in winter from 40 to 46 lbs. The instrument, therefore, shows not only that the road has been getting generally worse, but it points out the particular parts, and the exact amount of deterioration; thus enabling the proper authorities to say that the road has *become worse, the amount of the deterioration*, and the *exact part of the road where such deterioration has taken place.*

The public advantages to be derived from such a system of road inspection would probably be very great. It would show not only where the best plan of repairing roads has been followed, and point out where there are good and bad surveyors, but it would also show if the money of the trust is improperly applied or wasted on any line of road; and it will enable trustees, who let the repairs of their roads by contract, to determine whether or not the contractors have done their duty, and kept the road in the same state of repair as at first, or whether they had improved it, or suffered it to become defective.

There are many other uses to which the instrument may be applied, but the foregoing are the principal ones.

Specimen of the Manner in which it is proposed to keep a Registry or Journal of the State of Repairs of any Road.

FROM LONDON TO

First Quarter of First Mile.

DIST.	1829.		1830.		1831.		1832.		1833.		1834.	
Yards.	Summer.	Winter.	Summer.	Winter.	Summer.	Winter.	Summer.	Winter.	Summer.	Winter.	Summer.	Winter.
10	20	35	15	30	15	32	25	40				
20	30	35	25	30	27	32	35	40				
30	25	30	20	26	22	28	30	35				
40	28	33	21	28	24	30	35	40				
50	29	33	22	28	26	30	40	50				
60	35	39	26	29	30	35	45	60				
70	30	35	22	25	24	27	35	55				
80	30	36	23	26	25	28	40	45				
90	35	40	25	35	27	37	40	50				
100	40	43	30	36	32	38	45	55				
110	45	46	35	38	37	40	50	55				
120	50	55	40	45	42	47	60	65				
130	50	54	40	44	42	46	70	75				
140	50	55	40	46	42	48	55	75				
150	55	58	50	48	52	50	60	70				
160	52	66	43	41	45	45	65	56				
170	50	54	40	45	42	48	60	70				
180	51	55	46	45	48	47	60	75				
190	53	58	45	46	47	48	65	68				
200	55	60	50	52	52	55	65	70				
210	56	60	50	55	52	58	65	75				
220	55	60	45	55	50	60	65	75				
230	50	55	45	40	48	45	60	65				
240	50	55	45	35	47	37	55	75				
250	48	50	38	40	40	44	50	60				
260	45	50	35	40	38	45	55	65				
270	40	45	30	40	35	45	45	50				
280	40	45	36	40	40	45	50	60				
290	40	45	35	35	38	40	55	60				
300	46	50	36	40	40	44	50	60				
310	44	50	32	45	40	45	55	65				
320	43	48	31	45	35	50	50	60				
330	42	50	30	40	35	45	55	60				
340	40	46	30	40	35	45	50	60				
350	45	49	38	45	40	50	46	56				
360	50	55	40	45	45	50	55	65				
370	50	56	40	46	45	50	55	70				
380	51	58	40	48	44	49	60	70				
390	52	58	46	48	50	55	50	65				
400	53	56	40	46	45	55	60	70				
410	50	55	35	45	50	60	60	75				
420	50	54	36	40	50	70	80	85				
430	55	58	38	40	60	80	80	90				
440	50	58	38	40	80	88	85	95				
Total	1,958	2,176	1,567	1,776	1,783	2,046	2,361	2,780				
Horse power	44·5	49·45	35·6	40·36	40·52	46·5	53·6	63·18				

DESCRIPTION OF THE INSTRUMENT.

The framework is of wrought iron, about two feet six inches long, and eighteen inches wide. In this frame a dynamometer and brass cylinder are placed; the dynamometer is connected by its arm to one side of the frame, and the cylinder is secured in the frame by trunnions, which are cast on it, and which turn in a circular hoop or belt, firmly screwed to one side of the frame, and a bar running across it. The dynamometer, or weighing-machine, which forms part of the instrument, was introduced, some years ago, by Mr. Marriott; and as it is now so generally known, being used in mail-coach and other offices instead of the common steelyard, or scales requiring weights, it is needless to describe it here. On my applying the weighing-machine, in its simple form, to measure the draught of carriages, I found that the index vibrated so quickly, and over so large an arch of the circle, that it was impossible to observe the point indicating the force of draught; for a horse exerts his power by a succession of impulses, or strokes of his shoulders against the collar, at every step he makes, and not by a constant uniform pull, as is generally supposed. To remedy this inconvenience, and do away with the vibrations, I applied a piston, working in a cylinder full of oil, and connected with the dynamometer in such a manner that when any power or force is applied to it, so as to carry 'round the index, the piston is at the same time moved through the fluid. The connection of the dynamometer with the cylinder is by means of a lever working on a pivot; the arms of the lever are of unequal length; the tail-piece of the dynamometer is connected with the short arm, at a distance of two inches from the centre, or fulcrum, by means of a pivot-joint at precisely the same distance from the fulcrum; a flat bar of iron is connected with the longer arm, by a joint similar to that by which the tail-piece is connected with the short arm, so that any power or weight applied to the bar will produce the same effect on the index as if the power was applied directly to the tail-piece of the dynamometer; this bar passes over a friction roller, and to it the power of the horses is applied when in use, by means of traces and a bar, as in the ordinary mode of draught. At the extremity of the long arm, the piston-rod is connected by a joint similar to the others; the piston-rod, after passing through a stuffing-box in the cap of the cylinder, is screwed into a piston, or circular plate of thin brass perforated with small holes; and out of one part of the circumference a square notch is cut, the use of which will be hereafter described.

By this construction the resistance of the fluid to the piston, which acts at the extremity of the long arm of the lever, prevents its turning round the fulcrum to the extent it otherwise would do when it is acted upon by any sudden impulse applied to the bar; it will, however, move over a space proportioned to the intensity of the force applied; and if the pulls follow each other in rapid succession, the piston will move slowly out, and the index will turn round steadily and uniformly, until the power is balanced by the spring of the dynamometer, at which time the index will point out on the dial very nearly the weight or power which is equivalent to the draught.

The divisions on the dial-plate of the dynamometer decrease from zero upwards, in order to compensate for the increased force which the spring exerts in proportion as it is wound up: in consequence of this, the index does not pass over equal spaces when equal forces are applied in different states of tension of the spring; the piston, therefore, will not pass through equal spaces in the cylinder, and the vibrations would consequently be greater in the higher numbers, because, the velocity of the piston being less, its resistance through the fluid will be less, at the same time the power opposed to it is greater. To obviate this, and make the index equally steady on all parts of the dial, a narrow slip of brass, formed into an inclined plane, is sòldered to the inside of the cylinder, parallel to its axis, the largest part being at that end of the cylinder towards which the piston rises when the index moves towards the greater power.

The notch, which was before mentioned as cut in the side of the piston, exactly corresponds in size with the largest part of this inclined plane, so that when the piston is at the upper end of the cylinder, the notch is completely filled up by the inclined plane; on the contrary, when the piston is at the lower end of the cylinder, the notch is open: by this contrivance the aperture through which the fluid is obliged to pass, as the piston moves from the lower end of the cylinder to the higher, is gradually contracted, and, of course, the resistance of the piston through the fluid gradually increases, and compensates the increased power of the spring, rendering the vibrations nearly uniform from the lowest to the highest power. This compensation is analogous to that by which the fusee regulates and gives uniform power to the main-spring of a watch.

METHOD OF USING THE INSTRUMENT.

To preserve the instrument from warping, bending, or other injury, it is embedded in a solid block of elm, which can be screwed or clamped to any carriage; the swingle-tree is hooked into the eye of the draught-bar; the shafts or pole of the carriage may remain in their ordinary position, but care must be taken that no part of the moving power is communicated to the carriage, except through the agency of the instrument. The draught of a carriage over any portion of ground is ascertained as follows:—*

One assistant walks along the side of the carriage, and observes the weight, or force, shown by the index on the dial; at every step he calls out the numbers, which another assistant writes down in a book; these numbers are then added together, and the sum divided by the number of observations that have been made: the quotient will be the mean power, or draught, required to draw the carriage over that portion of the road. Thus, for instance, the instrument was fixed on the fore carriage of a common four-wheeled waggon, and two horses attached to it; it was then drawn over the pavement in Piccadilly, between the Duke of Devonshire's house and the corner of Dover Street; the numbers given by the index were, 50 lbs., 45 lbs., 50 lbs., 50 lbs., 55 lbs., 50 lbs., 45 lbs., 40 lbs, 45 lbs., 45 lbs., 50 lbs., 45 lbs., 50 lbs, 55 lbs.; the sum of these is 670 lbs., which, divided by 14, the number of observations, gives 48½ lbs. for the mean force which the horses must exert to draw the empty waggon over that part of Piccadilly pavement.

As the street in that part rises 1 foot in 156 feet, it is evident that the draught is greater than if the street was horizontal. To ascertain what it would be if it was horizontal, it is necessary to apply a correction to the draught actually shown by the instruments.

By theory we know that the power required to retain a carriage on an inclined plane bears the same proportion to the weight of the carriage and its load that the height of the inclined plane bears to its length; but as the mean velocity of the matter in the wheels of a carriage is dif-

* Since this was written, the instrument has been much improved by Mr. Macneill: it is now mounted in a light phaeton, and, besides marking the draught at every ten or twenty yards, it points out the distance run, and the rates of acclivity or declivity on every part of the road.

ferent from the velocity of the axis up the inclined plane, another correction would be necessary to get the motive power up the plane; and to persons not acquainted with the subject, the final result might appear doubtful. To alleviate every objection of this kind, I had a platform of timber erected, over which the same waggon that was used in the experiments on roads was drawn at different rates of inclination; and the power required to draw it up the inclined plane, with a uniform velocity of two miles and a half per hour, in each case, was carefully determined by dead weights passing over pulleys. By this means the correction for several rates of inclination was practically ascertained (without having recourse to theory or calculation), and the table of correction which I have inserted was formed for all slopes usually found on turnpike roads.

By this table, the correction for a slope of 1 in 156, is found to be 15 lbs.; hence the horizontal draught of a waggon over the paved surface between the Duke of Devonshire's house and Dover Street, will be 48¼, less 15¼, or 33 lbs.

In this way the surface of the whole road between London and Shrewsbury has been tried, the results of which are given in the following tables.

OBSERVATIONS ON THE USE OF THE INSTRUMENT.

A road with a smooth and uniform surface, which is merely preserved in that state by raking, appears to the generality of persons who travel over it to possess all the requisites of a good and perfect road, as such persons have no means of judging of the power necessary to draw carriages over it, or the exertion required from the horses; but by means of this instrument that power can be ascertained, and, consequently, the comparative merit of any line of road can be determined with absolute certainty, and the experiments made with the instrument will show how very important it is to the country to have the public roads constructed and maintained on true principles.

In some instances metal rails are laid on the sides of turnpike roads, with the same undulations and rates of inclination as the road; yet on these railroads a horse will usually perform as much work as five or six horses will do on the common road. This great difference in the useful effect of horses can alone be attributed to the friction of the road surface exceeding that of the metal rails; for the friction of the axles of the waggon will be nearly similar, and the resistance of gravity, arising from the inclinations, is, in this case, the same; hence the superiority of the one road over the other depends entirely on the surface.

The greatest resistance which a horse has to encounter, when in draught on turnpike roads, arises from gravity, which begins to act the moment the road ceases to be horizontal; and when the inclination exceeds one in thirty, which it often does, the additional power required is very great, as may be seen by the table of corrections; at the same time, the power of the horses is from the same cause much diminished. It is, therefore, the more necessary that the surface of the hills should be hard, solid, and composed of such materials as the wheels of carriages cannot

By making experiments with this instrument on every part of a turnpike road, both in summer and winter, and forming an exact table, showing the resistance of the surface, and the materials with which it is repaired, a complete register would be had of the state of the road; and any improvement or falling-off in the general management of the repairs of each part would be clearly perceptible, as also the amount of such improvement, or the reverse.

The following is an extract from the Appendix of this Report, containing Mr. Macneill's description of the experiments referred to: —

EXPLANATION OF TABLE, NO. 3.*

For the purpose of ascertaining the draught up different hills, with different velocities, the instrument was attached to a common stage coach, which weighed 18 cwt., exclusive of seven passengers. Stations were marked out on different parts of the road, of which the inclinations and the lengths were accurately determined, and the time of passing over each was ascertained by means of a stop-watch.

The results of these experiments are detailed in this table.

The first column contains the number of each experiment; the second, the rate of inclination of the hill; the third, the number of observations made on each; the fourth, the length of the hill or inclined plane, in feet; the fifth, the number of seconds in which the carriage was drawn up the hill; the sixth, the corresponding velocity in feet per second; the seventh, the velocity in miles per hour, calculated to the nearest quarter; and the eighth column contains the corresponding draughts, or force applied, in pounds.

Thus, in the first line of the first experiment, where the inclination of the hill was one in fifteen and a half, and the velocity three miles and a half per hour, the draught was 271 lbs.; and when the velocity was increased to twelve miles per hour, as shown in the fourth line of the

* Table No. 1. contains a detailed account of the experiments made on the Holyhead Road, and Table No. 2. the corrections for gravity according to the inclinations of the road.

same experiment, the draught was also increased from 271 to 325 lbs.

The part of the roads elected for these trials was of an uniform surface, the resistance of which was previously ascertained by drawing a waggon over it, to be an average between the worst and most improved parts of the Holyhead road; and although the velocities are not so varied, or so high as might be wished, yet several conclusions may be drawn from these experiments, of considerable importance in road engineering; one of which is, that the draught of a stage coach on a common turnpike road increases in a less ratio than the velocity increases, and not as the square of the velocity, which many persons have supposed, as is found to be the case in the steam carriage on a rail-road. From this it appears, that the resistance, arising from friction, of a steam carriage on a rail-road, and the resistance of a stage coach on a good turnpike road, are governed by the same laws of motion; and that whatever advantage may be gained by a quick transport of passengers, by means of a steam coach on the former, may also, probably, be attained by the same means on a well-made turnpike road.

TABLE III.

No. of Experiments.	Rates of Inclination.	No. of Observations.	Distance in Feet.	Time in Seconds.	Velocity in Feet per Second.	Velocity in Miles per Hour.	Draught of Stage Coach in lbs.
1	2	3	4	5	6	7	8
1	One in 15½	32	576	115″	5·0	3½	271
		22		111	5·2	3½	276
		13		47	12·2	8½	298
		13		33	17·4	12	325
2	One in 19	48	741	178	4·1	2¾	252
		17		82	9·0	6	290
		24		80	9·3	6¼	293
		32		66	11·2	7½	303
3	One in 20	35	750	136	5·5	3¾	253
		34		126	5·9	4	263
		16		81	9·2	6¼	272
		22		76	9·8	6¾	280
4	One in 21½	16	294	58	5·1	3½	237
		16		50	5·9	4	245
		8		33	8·9	6	258
		7		30	9·8	6¾	264
5	One in 23	29	522	99	5·3	3½	226
		23		78	6·7	4½	233
		13		55	9·5	6½	243
		13		49	10·7	7¼	250
6	One in 23½	45	1,032	189	5·5	3¾	230
		22		127	8·1	5½	240
		18		88	11·7	8	248
		—		83	12·4	8½	253
7	One in 26. Paved bottom, Hartshill stone surface.	16	387	69	5·6	3¾	200
		20		68	5·7	3¾	202
		13		42	9·2	6¼	215
		12		28	13·8	9½	223

TABLE III.—*continued.*

No. of Experiments.	Rates of Inclination.	No. of Observations.	Distance in Feet.	Time in Seconds.	Velocity in Feet per Second.	Velocity in Miles per Hour.	Draught of Stage Coach in lbs.
1	2	3	4	5	6	7	8
8	One in 26½. Not paved, limestone surface.	27 19 14 12	570	103″ 102 56 47	5·5 5·6 10·2 12·1	3¾ 3¾ 7 8¼	221 220 230 236
9	One in 28.	26 28 13 12	654	123 110 68 55	5·3 5·9 9·6 11·9	3½ 4 6½ 8	197 204 211 218
10	One in 30½.	15 12 9 6	300	58 52 30 23	5·2 5·8 10·0 13·0	3½ 4 6¾ 9	161 174 187 210
11	One in 33.	38 15 17 15	711	129 90 80 55	5·8 7·9 8·9 12·9	4 5½ 6 8¾	153 175 182 198
12	One in 34½. Patches of new stone, not worked in or consolidated.	30 11 10 13	534	95 65 55 43	5·6 8·2 9·7 12·4	3¾ 5½ 6½ 8½	186 196 200 214
13	One in 38½. Sub-pavement, surface quartz stone.	19 18 13 10	384	68 65 41 30	5·6 5·9 9·4 12·8	3¾ 4 6½ 8¾	146 150 167 170
14	One in 39. No sub-pavement, nine inches of limestone.	16 17 16 13	543	84 80 63 60	6·5 6·8 8·6 9·1	4½ 4¾ 5¾ 6¼	180 183 212 215

TABLE III.—*continued.*

No. of Experiments.	Rates of Inclination.	No. of Observations.	Distance in Feet.	Time in Seconds.	Velocity in Feet per Second.	Velocity in Miles per Hour.	Draught of Stage Coach in lbs.
1	2	3	4	5	6	7	8
15	One in 57. No sub-pavement, quartz stone surface.	21	552	97″	5·7	3¾	150
		12		85	6·5	4½	153
		13		66	8·3	5¾	160
		10		39	14·1	9½	168
16	One in 63½. No sub-pavement, six inches of lime-stone.	21	525	89	5·9	4	147
		24		88	6·0	4	147
		13		47	11·2	7½	182
		13		41	12·8	8¾	202
17	One in 118.	18	603	107	5·6	3¾	134
		26		91	6·6	4½	140
		11		60	10·0	6¾	146
		17		50	12·1	8¼	153
18	One in 137½.	35	741	122	6·1	4¼	133
		18		95	7·8	5¼	136
		18		75	9·9	6¾	140
		18		62	12·0	8¼	150
19	One in 156 rise.	44	861	161	5·4	3¾	128
		25		130	6·6	4½	133
		20		85	10·1	7	139
		22		61	14·1	9½	145
20	One in 156 fall.	41	861	205	4·2	2¾	82
		15		84	10·2	7	95
		13		57	15·1	10¼	100
		16		54	15·9	10¾	105
21	One in 245 rise.	28	648	105	6·2	4¼	125
		16		70	9·2	6¼	128
		15		48	13·5	9¼	131
		12		45	14·4	9¾	138

TABLE III.—*continued.*

No. of Experiments.	Rates of Inclination.	No. of Observations.	Distance in Feet.	Time in Seconds.	Velocity in Feet per Second.	Velocity in Miles per Hour.	Draught of Stage Coach in lbs.
1	2	3	4	5	6	7	8
22	One in 245 fall.	39 22 23 11	648	114″ 103 75 47	5·7 6·3 8·6 13·8	4 $4\frac{1}{4}$ 6 $9\frac{1}{2}$	96 100 107 117
23	One in 600 rise.	42 25 20 23	993	108 103 72 65	9·2 9·7 13·8 15·3	$6\frac{1}{4}$ $6\frac{1}{2}$ $9\frac{1}{2}$ $10\frac{1}{2}$	112 114 122 130
24	One in 600 fall.		993	177 147 100 64	5·6 6·8 10·0 15·5	$3\frac{3}{4}$ $4\frac{1}{2}$ $6\frac{3}{4}$ $10\frac{1}{2}$	100 110 115 127

If we suppose the power which a horse usually exerts in mail and fast stage coaches to be equivalent to a constant pull of 37 lbs.* over ten miles in a day, with a velocity of ten miles an hour, the effect will be equal to 651,200 lbs.; for $1760 \times 10 = 17{,}600$ yards in 10 miles, and this sum multiplied by 37 lbs. equals 651,200 lbs. drawn over one yard in the day; which number may be taken as a standard

* This is the power assigned by Mr. Tredgold, in his work on Railways, as that which a horse should exert working with a velocity of ten miles an hour, which, though sufficiently correct, as in the present instance, for a measure of comparison, should not be considered as a fixed standard of the power of a horse working at the velocity of ten miles an hour, as the formula which Mr. Tredgold has used appears to be founded on a limited number of experiments.

for horse power in comparing one line of road with another.

If, on this principle, we know the average draught over any line of road, and the length of that road in yards, we at once know the horse power to which it is equivalent, and, consequently, can compare it with any other line.

Thus, the average draught on the old road between Barnet and South Mims, multiplied by the length of that road in yards, is equal to 165,320; by the new line, the average draught multiplied by its length in yards is 139,028*; the difference between this and the old road is 26, 292, which, divided by 651,200, the power of one horse, gives ·04 part of a horse power; this multiplied by 500, the number of horses travelling over the road each day, and supposing each horse to be worth five shillings per day, the saving by making the new road will amount to 5*l.* per day, or to 1,800*l.* annually.

The result derived from this calculation depends not on any theory or abstruse calculation; it is a matter of *fact, and cannot be disputed;* and is, perhaps, one of the most useful practical applications of the Road Instrument; for without it the most refined and difficult application of algebra to the plans and sections of the above roads would only give an approximate value, as the state of the surface could not be brought into the calculation, *except by guess*, and this would be little better than judging by the sections, as heretofore practised, without any decided or fixed principle.

* See the annexed table of the actual draught on every twenty yards of this road.

BARNET AND SOUTH MIMS, OLD ROAD.

5th July 1833. *State of the Surface.* 3 miles, 352 yards.

	Going from Barnet to Mims.								Going from Mims to Barnet.							
Distance.	Draught.	Draught.	Draught.	Draught.	Draught.	Draught.	Draught.	Draught.	Draught.	Draught.	Draught.	Draught.	Draught.	Draught.	Draught.	Draught.
1	23	7	30	14	12	10	25	10	50	0	32	85	0	65	25	18
2	30	10	35	13	3	5	40	8	85	0	23	95	0	55	20	18
3	27	8	35	20	2	7	40	0	75	0	50	100	0	80	23	14
4	21	3	30	32	3	10	58	0	80	0	85	73	0	45	10	14
5	21	2	24	43	2	12	30	20	70	0	10	70	0	25	18	17
6	15	4	7	25	7	6	50	10	62	0	0	55	0	27	40	24
7	12	5	38	20	0	6	55	3	50	0	0	55	0	35	52	20
8	15	5	8	41	0	10	60	0	55	0	7	70	0	45	55	28
9	20	3	28	40	0	10	70	5	50	0	21	45	7	45	48	15
10	10	5	24	56	0	10	75	10	50	0	16	45	3	27	37	17
1 F.	19·4	5·2	25·9	30·4	2·9	8·6	50·3	6·6	62·7	0	24·4	69·3	1·0	44·9	32·8	18·5
11	20	6	30	48	0	25	90	25	35	10	15	40	12	10	28	18
12	16	10	0	52	0	14	60	24	30	0	18	30	10	16	30	18
13	20	10	0	45	0	10	70	10	30	0	17	20	10	32	32	20
14	23	0	0	50	0	15	100	0	35	0	12	25	12	50	30	20
15	12	0	0	45	0	30	80	0	40	0	10	45	15	40	30	17
16	13	4	5	48	0	47	50	0	50	12	20	55	15	21	30	22
17	16	5	18	60	0	60	60	0	25	18	20	50	20	17	18	19
18	22	20	10	75	0	30	50	0	30	15	25	40	15	37	30	12
19	23	17	0	75	0	0	50	0	65	25	20	45	33	27	27	22
20	20	30	0	80	0	0	65	0	73	15	35	55	30	40	20	20
2 F.	18·5	10·2	7·3	57·8	0	23·1	67·5	5·9	41·3	9·5	19·2	40·5	17·2	29·0	27·5	18·8

21	27	30	0	75	0	0	52		60	3	40	33	30	32	20	14
22	22	25	0	55	0	15	32		15	13	40	38	37	40	22	16
23	16	34	0	52	0	20	5		0	5	38	23	45	55	13	
24	22	17	0	50	0	16	0		0	7	40	10	55	57	20	
25	28	18	0	60	0	25	0		0	2	47	4	60	50	18	
26	26	17	0	85	3	13	0		0	3	50	8	75	30	13	
27	31	23	0	45	0	14	0		0	0	45	27	90	30	17	
28	25	22	0	4	0	13	0		0	10	40	45	80	40	15	
29	23	12	0	0	0	20	20		0	10	42	45	80	20	7	
30	20	14	0	10	0	22	13		0	25	50	40	95	24	8	
3 F.	24·0	21·2	0	43·6	0·3	15·8	12·2	6·25	7·5	7·8	43·2	27·3	64·7	37·8	15·3	3·0
31	20	10	0	12	0	12	45		0	33	55	25	95	22	10	
32	16	7	0	15	0	20	37		0	28	60	35	115	16	14	
33	21	4	0	10	0	35	45		8	20	40	40	98	18	10	
34	18	37	0	3	0	50	45		23	15	63	30	100	4	8	
35	17	28	0	5	0	25	50		55	12	60	7	80	5	18	
36	16	18	0	20	0	40	50		57	17	55	0	60	18	14	
37	12	7	15	24	0	45	55		50	15	85	0	65	20	18	
38	10	3	9	47	0	40	50		57	24	90	0	80	20	18	
39	4	4	16	20	3	40	65		60	20	75	0	70	22	12	
40	10	10	7	9	7	45	40		20	14	90	0	65	38	18	
4 F.	14·4	12·8	4·7	16·5	1·0	35·2	48·2		41·0	19·8	67·3	13·7	82·8	18·3	14·0	10·4
Mean of Half Miles.	19·07	12·35	9·47	37·07	1·05	20·67	44·42		55·62	92·7	38·52	37·7	22·55	32·5	22·4	
Mean of Miles.	15·71		23·27		10·86				32·44		38·11		27·50			

BARNET AND SOUTH MIMS, NEW ROAD.

4th July 1833. *State of the Surface.*

	Going from Barnet to Mims.								Going from Mims to Barnet.							
Distance.	Draught.	Draught.	Draught.	Draught.	Draught.	Draught.	Draught.	Draught.	Draught.	Draught.	Draught.	Draught.	Draught.	Draught.	Draught.	Draught.
1	12	20	0	25	5	40	17		70	18	17	40	52	40	42	
2	28	15	0	18	5	40	30		55	17	30	20	42	37	30	
3	18	20	0	12	22	45	20		42	15	25	17	45	43	40	
4	17	20	0	28	14	50	15		40	20	30	27	45	40	32	
5	23	20	0	33	10	33	15		25	35	40	13	45	35	38	
6	14	25	2	24	6	43	15		25	35	30	15	56	35	30	
7	18	18	4	20	12	37	12		25	22	25	8	52	40	24	
8	29	20	2	20	12	28	16		23	22	32	10	53	38	22	
9	20	15	0	28	20	25	10		25	25	40	18	38	35	20	
10	20	30	4	20	32	20	12		25	22	30	20	35	40	15	
1 F.	19·9	20·3	1·2	22·8	13·8	36·1	16·2		35·5	23·1	29·9	18·8	46·3	38·3	29·3	
1	28	22	2	20	20	20	6		30	17	40	13	35	40	10	
2	25	25	8	17	20	16	20		35	25	42	22	30	40	5	
3	15	22	6	25	23	16	16		45	19	50	17	30	30	7	
4	22	19	0	15	27	19	20		35	32	45	30	38	30	28	
5	20	15	1	30	25	12	18		26	30	50	25	40	20	4	
6	20	22	8	38	25	20	13		40	30	35	15	45	25	17	
7	16	15	5	44	32	22	10		30	25	27	18	35	20	15	
8	12	12	10	30	30	15	8		30	27	20	14	38	27	15	
9	14	30	1	32	30	30	7		30	12	30	20	50	23	18	
10	0	20	4	2	38	30	0			10	35	12	50	20	23	
2 F.	17·2	20·2	4·5	27·3	27·0	20·0	11·8		30·1	22·7	37·4	18·6	39·1	27·5	14·2	

1	0	18	3	20	28	40			25	6	25	10	47	14	
2	0	20	17	10	30	35			35	4	28	12	52	25	
3	0	14	6	10	30	30			33	15	20	18	62	30	
4	7	11	4	3	21	25			30	3	25	20	45	22	
5	8	8	2	10	25	15			19	12	27	20	55	30	
6	9	5	0	15	25	13			28	10	40	13	55	40	
7	12	6	2	20	23	20			20	10	40	12	50	57	
8	13	4	5	8	24	30			24	8	45	22	33	62	
9	13	3	4	12	30	23			25	2	38	21	30	65	
10	0	7	14	2	30	25			30	12	30	22	45	45	
3 F.	6·2	9·6	5·7	11·0	26·6	25·6			26·9	8·2	31·8	17·0	47·4	39·0	21·75
1	5	0	15	1	32	30			30	8	28	30	50	65	
2	0	1	17	13	35	28			30	12	30	30	42	65	
3	0	2	20	15	43	20			30	12	36	40	57	55	
4	0	0	22	6	51	30			45	10	47	35	65	40	
5	5	4	18	1	42	33			30	12	35	45	65	45	
6	5	0	30	14	48	36			55	20	40	40	34	35	
7	10	0	30	12	58	32			44	19	35	45	55	42	
8	10	0	20	12	50	28			38	12	33	58	42	38	
9	30	0	22	12	50	30			28	25	35	52	30	38	
10	13	0	20	12	50	30			25	20	43	60	43	45	
4 F.	7·8	0·7	21·4	9·8	45·9	29·7			35·5	15·0	36·2	43·5	48·3	46·8	
Mean of Half Miles.	12·78	12·70	8·20	17·72	28·32	27·85	Mean of Half Miles.		31·9	17·25	33·82	24·42	45·27	37·90	
Mean of Miles	12·74		12·96		28·08		Mean of Miles.		24·57		29·12		41·58		

Description of Mr. Macneill's new Instrument.

The instrument above described having been purchased by the Prussian government, Mr. Macneill laid before the Commissioners of Her Majesty's Woods and Forests a design for another, on an improved construction, which has been recently completed for them. The essential difference between this instrument and the former, consists in the self-registration of the effect of every pull or exertion that the horse may make in drawing the carriage, which renders it necessary to have an assistant to mark down the indications of the hand of the dynamometer, which was liable to several objections; for instance, it required some practice to be able to observe accurately the exact divisions marked by the hand, and to write them down with clearness and expedition; also it required that the observations should be made at equal intervals of time, as, for example, five or ten seconds, otherwise the mean power deduced from them would be incorrect. Inattention, therefore, on the part of the observer, was likely to produce an error in the result; and in any case it was difficult to observe and write down a series of observations taken at intervals of ten seconds, which should run on for four or five consecutive hours.

The new instrument, like the former, has a dial with an index, which shows the amount of power exerted at every step of the horse, the vibrations of the index being checked by means of a piston working in a cylinder of oil. But for the purpose of registering the pulls, and marking the power exerted by a horse at every part of the road, a long narrow sheet of paper is drawn, by means of cylinders, over a convex metallic surface or tablet, with a velocity proportioned to the speed of the carriage. The length of paper carried over the tablet is about two feet per mile of the road passed over by the wheels of the instrument. While the paper is thus drawn over the tablet, a pencil or metallic point, which is over the middle of the tablet, traverses the paper at right angles to the longitudinal motion. This point moves by the action of the horse, which pulls out the bar that acts against the spring of the dynamometer; but the direction of the motion of this bar being parallel to that of the paper, it is changed, by means of a system of wheels, into a direction at right angles to it, and also the pencil is thus carried across or at right angles to the line of direction of the bar.

The velocity or space passed over by the point is, by means of the same wheels, increased in the ratio of one to two nearly; thus, if the bar of the dynamometer be pulled out one inch, the point will be carried across the paper nearly two inches. Even the slightest pull on the spring is thus magnified and marked on the paper by the point.

The paper is ruled into faint blue parallel lines throughout its whole length; the intervals represent ten, twenty, thirty, forty lbs., &c., up to 600 lbs.; so that if the spring be acted upon by a power — say ten lbs., the point will be carried over the first interval, or to the line marking ten lbs; if equal to twenty or thirty lbs., the point will be carried to the lines marking twenty or thirty lbs. respectively. These lines will not be at equal distances from each other; for, as the power exerted upon the spring increases, the increments of that power will not be proportional to those of the space passed over by the bar and point; if the former be equal, the latter will diminish progressively, according to the form and strength of the spring; and therefore the distance between the lines must diminish in a certain ratio as the power increases. This ratio may be calculated, but, considering that the friction of all the moving parts must be taken into account, the simplest and perhaps most accurate method is to determine by actual experiment the spaces passed over by the point for every successive addition of weight or power applied to the bar of the dynamometer.

Besides registration of every pull exerted by the horse, the instrument also registers the levels and diversities of the road it passes over; this is effected by means of a heavy weight acting as a pendulum, the vibrations of which are checked by a piston working in a cylinder of oil, and forcing the fluid through the orifice of a tube on the outside of the cylinder which can be enlarged or contracted, so as to regulate the motion with great exactness, according to the nature of the surface passed over by the carriage; an even and smooth surface requiring a large passage for the oil, a rough surface, a contracted one. As the pendulum always hangs vertically, a rod connected with it will move backwards or forwards, according as the instrument ascends or descends; this motion of the rod being in the same direction, or parallel to that of the paper, it is changed, in the same manner as that of the pull, into one at right angles to it; and by this means a point is carried transversely across the paper, in precisely the same manner as the point which marks the pull. If the road be horizontal, the line described by the point will coincide with the zero line. If the road ascend, the bar is moved proportionally out, and consequently the point is carried to a certain distance from the zero or horizontal line, and will mark on the paper another line parallel to the zero line, so long as the acclivity continues the same; if that increases, the distance of the point from the zero line will also increase. Should the road descend, then the point will pass to the opposite side of the zero line, and by its distance from that line will show the declivity. The lines which mark the rates of clivity should be determined by experiment, as in the case of the tractive power; and this should be done every time the instrument is used, if the persons or load carried be different, or differently distributed.

In addition, there are also three dials, for registering the distance passed over. There is also a lever, by moving which a pin is forced into the paper, to enable the observer to mark the commencement or termination of any given distance, or the situation of any remarkable object on the road that he may wish to register. By means of another lever, the cylinders giving motion to the paper are thrown out of gear, and the instrument can travel without moving the paper. A very beautiful contrivance, designed by Mr. Renton, disengages the paper the moment each sheet is completed, without any attention on the part of the observer.

There are several other arrangements in the instrument different from the first one; but without a drawing it would be impossible to describe them to a general reader.

APPENDIX, No. II.

REPORT RESPECTING THE STREET PAVEMENTS, ETC. OF THE PARISH OF ST. GEORGE, HANOVER SQUARE.

1. PRESENT STATE.

In consequence of an application to me by the Pavement Committee of the inhabitants of this extensive parish, I examined the present state of the carriage-way and foot-path pavements, and endeavoured to learn the various circumstances connected therewith; I also made observations on the nature of the bottoming and shape of the stones.

The notorious imperfection of the carriage-way pavement having been the cause of this Report, it is needless to state that the surface is generally very uneven, and not unfrequently sunk into holes, so as to render it not only incommodious but dangerous to horses and wheel carriages.

The causes of this imperfection are various, and of an extensive and serious nature.

The stones, though generally of a tolerably good quality, are so irregular in their shape, that even their surfaces do not fit; they almost universally leave wide joints, and, instead of these joints being dressed square down from the surface, that is, at right angles with the face, they more frequently come only in contact near the upper edges, and, by tapering downward, in a wedge-like form, have their lower ends very narrow and irregular, leaving scarcely any flat base to bear weight.

This form also unavoidably leaves a great portion of space between the stones, which the workmen fill with loose mould or other soft matter of which the bed or sub-soil is composed.

Another great defect is caused by inattention to selecting and arranging the size of stones; they are but too commonly so mixed, that large and small surfaces are placed alongside of each other, and, acting unequally in support of pressure, create a continual jolting in wheel carriages, which, adding percussion to weight, is a powerful and destructive agent.

I must add to these defects another of an equally serious nature, that is, the imperfection of the bed on which the stones are placed.

This bed has, hitherto, but too generally been formed of very loose matter, easily convertible into mud; and this matter, instead of being compressed by artificial means, has unavoidably been loosened by a sharp-pointed instrument, to suit the irregular depth and narrow bottoms, and to fill the chasms between the joints of the paving-stones. From the width and irregularity of the joints, water easily sinks into, and converts the before-mentioned soft matter into mud, which, by the continual and violent action of carriage-wheels, is worked upon the surface, and leaves the stones unsupported.

This operation must be very evident to every person who reflects upon the sudden accumulation of mud upon the surface of the carriage-way pavement after a light rain, &c., or a continuance of soft weather.

This accumulation of the before-mentioned defects has, by degrees, arisen from carrying (perhaps well intentioned) economy to much too great an extent, and which has been accomplished by the easiest of all means, that is, promoting too indiscriminate a competition, and thereby reducing the price so unreasonably low, as to oblige contractors to procure inferior materials, and prevent them from bestowing the necessary portion of labour upon dressing and setting them.

There is a defect also respecting the management of contracts, in so far as to proceed by the almost unavoidable, and hitherto unchecked, mode of performing the work by the square yard of certain depths (say nine inches). Now, as I understand that the paving-stones are usually purchased by the contractors by weight, the more imperfect the shape is, the more profit he will have upon the superficial yard, which unavoidably must consist of a very considerable portion of loamy material, which is soon converted into mud.

The mode of repairing the carriage-way, if I am rightly informed, is equally imperfect, and has, no doubt, in a considerable degree, also arisen from the gradual introduction of low prices for repairing with old stones, by the square yard, however frequently repeated. This naturally produces hasty and imperfect workmanship.

The streets have likewise, of late years, been greatly disturbed by the laying down and repairing of water-pipes, &c. &c.

2. PROPOSED MODES OF CONSTRUCTING MORE PERFECT STREET CARRIAGE-WAYS.

The foregoing statement renders it sufficiently evident that the carriage-way pavements of the metropolis have reached a degree of imperfection which urgently demands reformation.

The defects having, however, been fairly stated, will assist us in discovering the best means of remedying them.

The result of the foregoing investigation undeniably is, that the surface of the carriage-ways of the streets is generally very uneven, not unfrequently dangerously rugged, and in constant need of repairs, and these combined circumstances have created strong prejudices against pavements.

Sundry modes have been proposed to get rid of these

very general and well-founded complaints, which I shall now proceed to discuss.

One of the boldest of these projects has not only been proposed, but actually, to a certain extent, put in practice, by making a total change from a pavement surface to that of small broken stones. This radical change appearing to me to require all the judgment and experience which can be brought to bear upon it, I have not only exerted myself personally to acquire information, but have submitted the subject to repeated discussions at sittings of the Civil Engineers' Institution, when numerously attended by many of the ablest and most experienced engineers and surveyors, not only of the metropolis, but of various parts of the kingdom.

The result of these able and very candid discussions was, an unanimous resolution that whin or granite pavement, of proper form and depth, laid on a sound bottom, is preferable to any other mode for carriage-ways for the metropolis and other large cities, in order to form a body of strength adequate to bear the pressure and shocks of innumerable carriages, many of them conveying several tons.

The chief objections advanced to small broken stone were as follows: — that they cannot resist the pressure caused by a very great intercourse, being liable to be thereby crushed and ground into dust, easily converted into mud; that this hasty and continual destruction and renewal would, in a great city, prove intolerably troublesome and expensive, while the dust in dry weather, and the mud in wet, would greatly incommode the intercourse in the streets, also private dwellings and public shops. Cases were instanced where absolute nuisances had been created by employing broken stones; and that it was well known that, in some large cities, the want of pavements led to accumulations of filth, very injurious to the health of the inhabitants. It was observed, as a constant and abundant supply of broken stones would be required for repairs always hastily performed while the streets were empty, that receptacles, such as made in country roads, would not readily be found in London, where space is so valuable and so fully occupied. And it was further observed, that a surface of broken stones frequently covered with dust and mud was more injurious to the feet of horses than a properly constructed pavement, which is also much easier for their labour. And, lastly, that the expense of making and maintaining a street carriage-way with broken stones, including the constant labour and carting away scrapings to different depositaries, would be at least fifty per cent. more than by a proper pavement.

These observations corresponding with my own sentiments and experience, I am led to recommend pavement in preference to broken stones for the carriage-way of the streets in St. George's parish.

3. OF THE BEST MODE OF CONSTRUCTING PAVEMENTS.

To obtain a smooth and durable pavement surface, the following essential matters must be attended to: —

1st. The bed, or bottoming upon which the stones are to be placed.

2dly. The quality, size, and shape of the stones.

3dly. The mode of contract for constructing and keeping the pavement in a state of repair.

1. *Bottoming.*

After the space between the foot pavements has been brought into a form, consisting of a very slight curve in the cross section, every devisable means should be resorted to in order to render it compact and solid. Where practicable, it will be advisable to have wheel carriages to run for some time over it, or, occasionally, water; or to use the roller and stamper. These operations performed, it is necessary to cover the whole surface with a stratum, or layer, of some sort of substance, which will effectually cut off all connection between the subsoil and bottom of the paving-stones. This must itself be indissoluble in water, and prevent any of the substratum from rising in the shape of mud. Where stone can be cheaply procured, a bed of it, broken very small, would perfectly answer the purpose; and hence it has been observed, that the present broken stone experiments, in certain streets, will not be an entire loss, because they will, at all rates, constitute a good bed for a proper pavement. But, as relates to the metropolis generally, I am persuaded that a bed of cleansed river ballast, about six inches in thickness upon an average, will be found to answer the purpose, and is to be obtained at a comparatively moderate expense.* This should also be rendered compact and solid; it might be travelled upon for some time without inconveniency, particularly in summer, being the season when paving is usually performed.

2. *Quality, Size, and Shape of Stones.*

The qualities of the granites hitherto used are not so materially different as to require much discussion; but as there are differences in stones from all quarters, the judgment of the surveyor, who has charge of the works, and is supposed qualified, ought to be constantly exercised to ensure a due fulfilment of the contract in regard to the materials. I have understood that by former and recent experiments, the Guernsey stone is of great compactness and durability. This deserves attention and a fair and impartial trial. The only objection I foresee, is in its disposition to smoothness of surface.

* On reconsidering this subject, I am of opinion that this quantity of ballast will not make a sufficiently strong bottoming, and that nothing short of twelve inches of broken stones, put on in layers of four inches each, and then completely consolidated by carriages passing over them, will answer the purpose. —T. Telford, *July* 18, 1833.

With regard to size, it ought to be regulated, in some measure, by the nature and quantity of the intercourse through the several streets: they may be conveniently divided into three classes. For streets of the first class, or greatest thoroughfare, the stones should be not less than ten inches in depth, from eleven to thirteen inches in length, and from six to seven inches and a half in breadth on the face.

For the streets of the second class, the stones should not be less than nine inches in depth, from nine to twelve inches in length, and five to seven inches in breadth on the face.

For the streets of the third class, the stones to be not less than seven to eight inches in depth, from seven to eleven inches in length, and four and a half to six inches in breadth on the face. Crossings ten to twelve inches in length, seven to eight inches in breadth; the depth to be according to the classes.

All these stones to be worked flat on the face, and straight and square on all the sides, so as to joint close, and preserve the bed or base, as nearly as practicable, of an equal size to the face; and stones of equal breadth on the face, must be carefully placed adjacent to one another. The inferior streets, mews, and passages may be paved with the inferior stones from the other three classes, and those stones unfit for any pavement may be usefully employed by being broken small, as bottoming for the pavement of the first class of streets.

With regard to the shape of the stones, those modes I have hitherto been considering have been supposed rectangular, with joints made exactly to fit close to each other, and which, if perfectly executed, taking into consideration all the angles, the strongest possible, and also the most simple, whether we regard the preparation in quarrying and dressing, or the practical operations on the streets—right angles admitting of a variety of size, but always fitting, however applied, and, of course, under all these circumstances, the cheapest.

3. *Modes of Contract, &c.*

In constructing new pavements by the present practice, I have already stated, that it is the interest of the contractor to work with stones of a defective shape. The making the superficial yard of face-work must, I conceive, still be continued as the rule; but, along with that, weight ought certainly to be combined, as a proof that the quantity of stone intended is really obtained; the shape of the stones must be accurately defined in a specification, and, *above all*, the surveyor or inspector must, by unremitting attention, see that every part of the contract is fully and faithfully performed. A further security for the perfection of the work would be obtained by making it a part of the agreement, that the contractor should keep the work in a perfect state of repair, at a given rate per yard, for a certain number of years; the necessary repairs to be from time to time pointed out by the surveyor, under the direction of the Committee.

In repairing the streets, as far as regards the stones now in use, although these stones are, undeniably, very imperfect, yet the quantity and value being so great, no project for rashly disposing of them is admissible; but a thorough improvement may be gradually accomplished in the following manner: —

The streets to be divided into three different classes. For the first, and most important, perhaps very few of the present stones are suitable; if there should be any, they may be reworked and replaced, and new stones of a proper shape (of course) provided for the rest of the street. The stones rejected from the first class should be carefully sorted and reworked, for the repairs of such of the other classes as they are fit for, taking care that the stones, in all cases, are worked into proper forms, as regards the joints and bottoms; that the bottoming or bed has been formed of proper materials, not convertible into mud by the water running down the joints, and that the stones, as to sizes, have been judiciously arranged.

It is now, I understand, the mode to repair by the yard superficial, in partial spaces pointed out by the surveyor, at a price per yard for each time. The contractor, therefore, has no inducement to have the operation performed in a complete and substantial manner; but, on the contrary, it is his interest to have a great quantity done by his workmen in a short time, because, the more frequently it fails, the more demand there is for his services. I do not by this insinuate unfair practices against any individual contractor, but the practice is undeniable.

To correct this apparent evil, it seems advisable to let a whole street, or certain number of streets, for a certain time, at a fixed price per yard; the necessary repairs to be pointed out by the surveyor, under the direction of the Committee.

From what I have here stated, it must be quite evident that, to acquire a necessary degree of perfection, the most unremitting and strict attention, on the part of the surveyor, is absolutely necessary; and that St. George's extensive parish is quite sufficient to employ the whole time of the most active and persevering man. He ought to have no other object, and his remuneration should be sufficient to attach him to his duty.

But even his most judicious and faithful exertions will be unavailing, unless a price is allowed equal to the fair value of the materials and workmanship, and a reasonable profit to the contractor, as I cannot help again stating, that the injudicious practice, which has of late years very generally prevailed, of reducing prices too low, has led to the imperfect condition the street pavements are now in, and which, in works of this nature, is a very mistaken economy; for of all things, streets of great thoroughfare should not only be commodious, but very seldom interrupted.

A very perfectly constructed pavement might, I am convinced, combine smoothness, durability, and, in the course of a few years, true economy.

WATER-PIPES.

Formerly, when the main pipes consisted of wood, the rapid decay, and consequently frequent repairs, created a constant interference with, and very considerable injury to, the pavements; but since the introduction of iron pipes, these inconveniences have been greatly lessened, as experience has proved that very few repairs are found necessary.

The failures which now take place are almost wholly in the lead service-pipes from the subsidiary mains to the cisterns. There are two causes for this: the one is, by the subsoil of the street decomposing the pipe; the other, which is the more serious, arises from the effect of frost. It would, therefore, be very desirable if some other material could be used. I am convinced that a small iron pipe may be substituted with advantage. It would withstand the pressure, and might be laid with an inclination, so as to discharge its water either into the subsidiary main or the cistern, and thereby prevent its bursting by the effects of frost; and, in most cases, it would afford a supply of water to the houses even in severe frosts, because these subsidiary mains are generally laid sufficiently below the reach of frost.

SEWERS.

With regard to the openings which must unavoidably be made for common sewers, they so seldom occur, and may be so effectually secured, that their consequences do not require any serious notice.

STREET-WATERING.

The present mode of watering the streets from the plug-pipes, by throwing it about with scoops, is extremely injurious to the pavement joints. I therefore recommend that water-carts be substituted.

FOOT PAVEMENTS.

In the course of my examinations, I found that the foot pavements, though in some few instances requiring repairs, and even improvement, yet were, upon the whole, in a very fair state, and did not seem to require any separate discussion, or material change from the mode now practised.

I may merely observe, that, in future, it will be an improvement, that its surface may have a gentle declivity from the areas, or houses, towards the street; and that the new stones should, in the principal streets, be good Yorkshire pavement, not less than four inches in thickness.

GENERAL OBSERVATIONS.

From the foregoing discussion it is evident,—

1. That pavement is the most advisable mode of constructing street carriage-ways.

2. That the present street pavement is very imperfect, because the bottoming to receive it, being composed of a variety of loose materials, all easily convertible into mud, is very unfit for a foundation. And the irregular depth and shape of the stones placed upon such a bottom, with a generally small base, having only the support arising from the friction of the very trifling breadth which comes in contact near the upper edges (even when these parts of the joints are good), are easily pressed down into the aforesaid loose materials; and although ramming may assist the present mode, yet the irregularity of depth and

base prevents it from constituting a perfect and durable pavement.

3. That it has become absolutely necessary to abandon the present imperfect mode of paving and repairing, and adopt a better one.

4. In order to commence the improvements (which I have stated may be gradually accomplished), I recommend that, in one of the streets of the first class, say about 100 yards, which are now in the most imperfect state, should be wholly taken up, and restored with new stones of proper dimensions and shape, placed upon well-prepared substantial bottoming, and the whole managed in the way described in the foregoing Report.—The stones taken up may be selected and re-dressed for the second class of streets.

For each operation, precise specifications, and drawings of the cross sections, should be prepared, and there should, in the commencement, be small specimens made of the new pavement, and also of the re-dressing, under the direction of a properly qualified person, who would enter with zeal into the spirit of the improvement. Upon these specifications and specimens, contracts might be entered into.

5. That, to ensure an effectual performance of what has been recommended, unremitting attention is absolutely required, and quite sufficient to occupy the whole time of an able, intelligent, experienced man.

I cannot close this Report without acknowledging the aid I have derived from the ready and full information afforded by the Agent of the Grand Junction Water Works Company in all that regards water-pipes, &c. &c.

(Signed) THOMAS TELFORD.

June, 1824.

APPENDIX, No. III.

OBSERVATIONS COMMUNICATED TO THE AUTHOR, ON MR. WALKER'S EVIDENCE BEFORE THE SELECT COMMITTEE OF THE HOUSE OF COMMONS, IN 1819.

When Mr. Walker gave evidence before the Select Committee on roads, in 1819, he delivered a section of the Commercial Road paving, and also of the East India road, both of which were printed.

In the same evidence he stated that, for heavy traffic in each direction, it would be an improvement to form the paving on the sides of the road, and leave the middle for light carriages, by which the carmen, when on the footpaths or sides of the road, could be close to their horses, without interrupting or being in danger of accidents from light carriages, and the unpaved part, being in the middle or highest part of the road, would be more easily kept in repair. He delivered a section according to this plan.

In 1820, two miles of the Middlesex and Essex road, from Whitechapel to Bow, were paved on the above plan. The specification for the contract was prepared by Mr. Walker, and the first part done under his direction. He appears to have taken great pains to ascertain by experiment the proportional durability of different stones; for which purpose he had pieces of equal weight and rubbing surface of Guernsey, Aberdeen, and Peterhead, placed loosely in a frame divided into squares, each stone occupying a square. The frame was moved back and forward for several days on a large block of stone with sand and water between, and the loss of weight noted each day. The Guernsey stood best, and the contract was made for it. This paving has now been in use thirteen years, with almost the heaviest traffic out of London on it, and, except the first year, when the contractor had to keep it up under his agreement, it has cost very little for repairs. It is now in excellent order, and the stones do not appear worn in the smallest degree. The paving is 9 feet wide, with a curb on each side, and the gravelled road between about 32 feet on the average; making the width of the road, exclusive of the footpaths, 50 feet.

Much praise for the excellent way in which this great road has been improved and kept, is due to the chairman of the trustees, John Henry Pelly, Esq. From being at one time the worst, it has, for twenty years, been the best, and is still one of the best roads out of London, and at the same time much the cheapest in point of toll. The tolls allowed by the Act are as under: those actually charged are 25 per cent. lower.

	s.	*d.*
For a horse, &c. laden or unladen - -	0	1½
For a coach, &c. with five or more horses -	1	4
For ditto, with three or four horses - -	1	0
For ditto, with two horses - - -	0	6
For ditto, with one horse - - -	0	4½
For two-horse cart with wheels from six to nine inches - - - - -	0	8
For waggon with two or more horses, and wheels less than six inches in breadth - -	1	6
For one-horse cart - - - -	0	8
For dray with one or more horses - -	1	0
For waggon with one or more horses, and wheels six to nine inches broad - - -	1	0
For ditto, if wheels more than nine inches -	1	6
Oxen, &c. - - - - -	0	10
Calves, sheep, &c., per score - - -	0	5

A payment at Mile End gate clears from London to Brentwood (18 miles), and tolls are due only once a day.

The formation of the tramway on the Commercial Road gave Mr. Walker another opportunity of proving the absolute wear and comparative hardness of granites. The experiments were made on pieces of the tram-stone 18 inches wide and a foot deep, which, after being accurately weighed, were laid down in one of the toll-gateways where all the traffic from the East and West India Docks uniformformly passed; on being taken up, the stones were again weighed, and the results were as under: —

First Series.

Laid, November 1828. Lifted, March 1829.	Time, Four Months.	
Description.	Loss of Depth.	
	Absolute.	Relative.
	Inch.	
Guernsey - - - - -	·028	1·000
Budle * - - - - -	·029	1·025
Herm † - - - -	·034	1·194
Blue Peterhead - - - - -	·051	1·809
Red Aberdeen - - - - -	·074	2·601
Blue Aberdeen - - - - -	·086	3·039

Second Series.

Laid, March 1830, Lifted, August 1831.	Time, Seventeen Months.	
Description.	Loss of Depth.	
	Absolute.	Relative.
	Inch.	
Guernsey - - - -	·060	1·000
Herm - - - - -	·075	1·190
Budle - - - - - -	·082	1·316
Blue Peterhead - - -	·131	2·080
Heyton - - - - -	·141	2·238
Red Aberdeen - - - - -	·159	2·524
Dartmoor - - - - -	·207	3·285
Blue Aberdeen - - .	·225	3·571

* Budle is a whinstone from Northumberland; the others are all granites.

† Herm is an island adjoining Guernsey.

The second of the above series may be considered the more correct, from the longer duration of the experiment, and from the dressing of the stones of the different descriptions at the time of the first series not being quite equal.

Remarks on Horse Railways, and Tramways, by John Macneill, Esq.

Railways worked by horses are becoming very general; some have been constructed both in Scotland, England, and Wales, and others are about being formed in Ireland to a very considerable extent. These lines, though worked by horses, have edge rails similar to the best description of locomotive railways, but of less dimensions; they should therefore be called horse railways, not tramways, which, as before stated, are constructed in a very different manner. A railway worked by horses possesses many advantages; in the first place, a horse will be able to perform nearly ten times as much work as he could on a common road, and the expense of construction is so moderate, when compared with locomotive railways, that they may be introduced into districts and countries where the trade and commerce would not justify such an expenditure, even if a sufficient capital could be raised for the purpose. They are certainly more adapted for the carriage of goods and agricultural produce than for the carriage of passengers where a very speedy transit is required; but where 9 or 10 miles an hour are sufficient, they are admirably adapted for a passenger trade, as proved on the Edinburgh and Dalkeith railway, and several others, and at that rate of velocity will probably be found to be the most economical mode of conveying passengers that has yet been introduced. The first cost of such railways will be found to vary from 3,000*l.* to 5,000*l.* per mile, according to the nature of the country through which they are carried; the total mass moved in a train of waggons drawn by one horse being so small as compared with the mass moved by a locomotive engine, that the increase of resistance, arising from gravity on an acclivity, bears but a small proportion to the increased resistance arising from the same cause on a locomotive line, and is therefore more within the compass of the moving power. The rates of acclivity may therefore be much greater on the former than on the latter description of railway; and the cuttings and embankments, which are the most expensive part of railways, may therefore be much less, and the cost of construction consequently reduced.

The rails may also be of much less weight per yard; on locomotive lines the rails should weigh from 60 to 70 lbs. per yard; as they have to carry the weight of the engine, which is often made to weigh from 8 to 12 tons, and the rails must be proportionably strong, though the waggons which carry the goods or passengers would not require rails of half these dimensions.

The wear and tear on the horse railway will be much less than on the locomotive line, for the power which propels the waggons in the former case does not touch the rails, or do any injury to them; whilst, on the contrary, the propelling power on the locomotive railway causes the greatest amount of annual repair and expenses. The effect which a horse can produce on a railway of this description will be seen from the following table.

——	Rates of Acclivity.	Number of Tons gross at 3 Miles per Hour.*	Number of Tons gross at 10 Miles per Hour.	Number of Passengers at 10 Miles per Hour.
One Horse	1 in 100	4½	2	15
	1 in 200	6¾	3	20
	1 in 300	8	3¾	26
	1 in 400	9	4	30
	1 in 500	9½	4½	37
	Horizontal	13	6¼	45

* The power assumed for a horse moving at the rate of three miles per hour is 150 lbs., and 70 when moving at the rate of ten miles an hour, and working short lengths, as in mail coaches.

Tramways, or what should more properly be called trackways, formed of flat metal plates without ledges, or of blocks of granite, have been used with considerable advantage, and have of late been introduced with considerable success. These trackways are more applicable to general traffic in the streets of towns, and on quays, wharfs, &c. than the edge rail, or tramway; but where they are likely to become most useful is on the sides of the present turnpike roads, where they may be adopted with the greatest public advantage.

The trackway in Glasgow is composed of plates of cast metal laid on stone supports, and flush with the surface of the road. They are about two inches thick, three feet long, and eight inches wide. They are laid on a part of the street which rises 1 in 20, yet on this steep incline horses drew up 4 tons with apparent ease. The expense of keeping this road in repair is very trifling.

Trackways of granite have been lately laid on the turnpike road between Coventry and Nuneaton, and are found to answer exceedingly well. In this case, as the horses work two abreast, they are obliged to travel on the stone tracks, as they work in the lines of the wheels; in the others already mentioned, the carts or waggons are drawn only by one horse in the shafts, by which means he travels on the pavement between the lines, and not on the tracks themselves. This, however, produces no inconvenience whatever, and no obstruction to the horses from slipping or any other cause. On this trackway the blocks are of granite 12 in. × 9 in. and 3½ feet long; the space between the lines is not paved. The work cost at the rate of 1,519*l.* per mile.

A trackway is about to be laid down on part of the Holyhead road, the specification for which is given in another part of this work, and when finished will perhaps be the most perfect one of the kind hitherto made. That part of the road on which it is to be constructed is generally very soft, and difficult to draw over, arising from the nature of the soil and the limestone used in repairing it. The power required to draw one ton over a horizontal part of it was found by experiment to be 150 lbs., and on the acclivity of 1 in 20, 294 lbs. When the trackway is laid the horizontal draught will be reduced to 13 lbs., and the draught on the acclivity of 1 in 20 to 131 lbs. This is equivalent to reducing the acclivity from 1 in 20 to a horizontal, supposing the surface of the road to remain as it was when the experiment for ascertaining the force of traction was made. As the part of the country where this work is to be executed is a considerable distance from granite quarries, the work will be expensive; perhaps as much so as in any part of England. The expense of similar works where granite could be procured at a moderate rate would be much less, as will be seen by the following detailed estimate of one running yard of the proposed work.

	£	s.	d.
Getting out earthwork and drains - -	0	1	0
Drains at each side - - -	0	1	6
1¼ yard sandstone and coating, at 1*s.* 6*d.* -	0	1	10½
1¼ yard of limestone broken, at 3*s.* - -	0	3	9
¼ yard of Huntshill broken, at 8*s.* - -	0	2	0
Lime and sand - - - -	0	1	0
Blocks of granite, at 3*s.* 9*d.* per cubic foot -	1	6	3
Curb-stones - - - -	0	4	0
Workmanship, pitching, grouting, ramming, laying blocks and curbs - - -	0	2	3
Total expense of one running yard - -	2	3	7½

APPENDIX, No. IV.

REPORT OF THE SELECT COMMITTEE APPOINTED TO INQUIRE INTO THE AMOUNT OF ALL SUMS OF MONEY RECEIVED, EXPENDED, AND REPAID BY THE COMMISSIONERS FOR THE IMPROVEMENT OF THE HOLYHEAD AND LIVERPOOL ROADS.

May 30, 1830.

THE Committee have proceeded in making the inquiries referred to them, and find from the accounts which have been laid before them, that the Commissioners appointed by the act of the 55th Geo. 3. c. 152. for the improvement of the Holyhead road, have received the sum of 759,718*l.* 6*s.* 11*d.*; that of this sum 338,518*l.* 14*s.* 1*d.* has been granted by Parliament, without any condition for repayment, for works in North Wales; that 394,114*l.* 6*s.* 6*d.* has been granted by Parliament, or advanced by the Exchequer Bill Loan Commissioners, by way of loan, towards the building of the Menai and Conway bridges, the making of the new road across the Island of Anglesey, and the improvement of the road between London and Shrewsbury; and that repayments have been made to the amount of 103,633*l.* 2*s.* 2*d.* The details of the expenditure of the sum received by the Commissioners, of 759,710*l.* 6*s.* 11*d.*, are stated in an account in the Appendix.

In order to ascertain in what manner the Commissioners have applied the money entrusted to them, the Committee examined their engineer, Mr. Telford, who produced the statement which is here inserted, containing a short description of each work that has been executed by the Commissioners since they were appointed to superintend the improvement of the Holyhead road, in the year 1815, under the following heads: —

1. Roads made in North Wales on the London and Holyhead Mail line.

2. Roads made in North Wales on the Chester and Holyhead Mail line.

3. Embankments on the Stanley Sands and at Conway.

4. Bridges over the Menai Strait, and over the river Conway.

5. Roads made between London and North Wales on the London and Holyhead Mail line.

6. The Harbours of Holyhead and Howth.

7. The Road from Howth to Dublin.

8. The widening and deepening the Channel through the Swilly Rocks in the Menai Straits.

Statement of the Works performed between the Years 1815 *and* 1830.

HOLYHEAD ROAD. — NORTH WALES.

Holyhead to Chirk.

Between Holyhead Pier and Chirk Bridge is a distance of 83 miles 1,320 yards. The whole may be fairly considered a new road, as the short pieces of the old road were entirely re-made. The whole of the roadway is constructed with a substantial rubble-stone pavement, carefully hand-set, and covered with a six-inch coating of properly broken stone. There are, in all cases where found necessary, breast and retaining walls of stone, with numerous side and cross drains, all constructed in the most perfect manner. The whole is protected with stone walls; those upon precipices built with lime mortar, most of the others pointed with it. There are several considerable bridges, also numerous cuttings and embankments, in that mountainous country; one, particularly, at the village of Chirk, is 50 feet in height. Four miles of branch roads have been made. The toll-houses and gates are on a new construction, as are the milestones; and sufficient recessed depôts for stones have been made in every part of the road. An entirely new inn has been built in the middle of the island of Anglesey, upon the new line of road.

CHESTER LINE.

Tally Pont Hill.

From Tally Pont Bridge a new road has been made upon comparatively level ground, to avoid the inconveniently steep inclination of Tally Pont Hill, and to save distance, 1 mile 132 yards. This road is made with a pavement bottom, and a covering of hard field pebbles, properly broken.

Penman Mawr.

This improvement consists chiefly of rock cutting, in some parts 30 feet in height, with high breast and retaining walls, stone parapets laid in lime mortar; the roadway is formed of pavement bottoming and a coating of broken stone; so that this formerly frightful precipice is now a safe trotting road.—Distance improved, 1 mile 231 yards.

Penman Back.

Between Penman Mawr and the town of Conway, excepting a short distance, a new road has been made to avoid Sychnant Hill, over which the mail road formerly passed, by dangerous inclinations, to upwards of 540 feet above the level of the sea. The new road is nearly level, having no inclination more than one in twenty-five, and that only for a short distance of about 100 yards; the rock cutting required in some instances, in the face of a precipice, 100 feet in height. The roadway is, in all respects, similar to that at Penman Mawr, as to breast and retaining walls, parapets, &c. — Distance improved, 4 miles, 1408 yards.

Rhyalt Hill.

Up a valley between St. Asaph and Holywell, a road has been made to ease the steep and long-continued ascent of Rhyalt Hill, which, in some cases, was one in seven. The new roadway is made with bottom paving and broken limestone, with very good side and cross drains. — 2 miles 1166 yards.

EMBANKMENTS IN NORTH WALES.

Stanley Sands.

Near Holyhead there is an inlet of the sea, known by the name of the Stanley Sands: over this estuary an embankment, 1144 yards in length, has been made; the height above the undisturbed surface of the sands, in the middle, is 29 feet: the breadth at the top, including the parapet walls and outer facing, is 34 feet; the slopes on each side are at the rate of three to one, and these slopes are faced with rubble stone, two feet in thickness; on each side of the road there is a parapet four feet in height, coped with cut stone. The roadway is 24 feet in width; it has a paved bottom and a coating of broken stone. In order to admit the tide to flow into the space on the west side of the embankment, there is a bridge built upon the only piece of natural rock found in that part of the estuary. This work was executed in two years; 156,271 cubic yards of earth, and 25,754 cubic yards of rubble stone, were deposited in forming it: it has been completed seven years, and is now in a perfect state.

Conway Embankment.

The eastern approach to Conway Bridge is formed by an embankment upon the sands, over which the tide usually flowed, and rendered it a very difficult and dangerous passage. The distance from the eastern shore to the island is 672 yards; the height of the embankment, on account of the sand being swept away by the violence of the tides during the execution of the work, is 54 feet; its breadth at the base is 300 feet, and 30 feet at the road-way: the side slopes are faced with rubble stone.— 261,381 cubic yards of earth, and 51,066 cubic yards of rubble stone, were employed in forming it. The whole has been finished three years, and is now in a perfect state.

BRIDGES IN NORTH WALES.

Besides several stone bridges, three of a novel description were required:—

Menai Bridge.

Over the Menai Strait, which separates the island of Anglesey from Carnarvonshire, in order to supersede an inconvenient ferry, it was found, after many years investigation and discussion, that in a navigable and rapid tide-way, a bridge upon the principle of suspension was the most practicable and economical; a bridge of that description, therefore, was begun in 1818, and successfully completed and opened on the 30th of January, 1826. This structure being of very unprecedented novelty and magnitude, considerable apprehensions were entertained concerning its stability; the engineer, therefore, by the advice of his friend, the President of the Royal Society (one of the Commissioners), considerably increased the height of the piers, and the dimensions of the masonry and ironwork, beyond the original design, and this unavoidably led to corresponding increase of expense; but as all the works were paid for at the prices previously fixed in making the first estimate, and as the quantities have been ascertained by measurements and weights correctly made by the resident engineer, the public has only paid for what was actually found in the work, and the edifice is thereby rendered more substantial.

The contractor for the iron works having made a claim on the Commissioners for alleged loss sustained by him in consequence of the unprecedented rise in the price of iron, the Commissioners felt themselves justified, on inquiry, in representing to the Treasury that the difference between the price paid by him for 2,000 tons of iron, employed on this and the Conway bridges, and the price at which the contract had been made, exceeded 4,500*l.*; but this claim was not admitted.

The distance between the points of suspension, for the middle opening, is 580 feet, and between the pyramids and toll-houses about half as much, to which is to be added what passes down the galleries to the places of fixture in the rocks, making the whole length of each main chain 1750 feet, or one third of a mile.

The height from low water to the top of the saddles on the pyramids is 181 feet; and between the saddles and the roadway, 60 feet.

The breadth of the platform is 30 feet, and consists of two driving ways and a footpath between them.

There are four stone arches on the Anglesey side, and three on the Carnarvonshire side, each 52 feet 6 inches span.

This bridge has been in constant use four years, has required no repair but painting, and is now in a perfect state.

Conway Bridge.

At the town of Conway, between the before-mentioned island and the rocks in front of the old castle, there is a space through which the tide flows with very considerable velocity: over this space there has been made a bridge on the same principle as the Menai; it is 327 feet between the points of suspension; in this there is only a single roadway. The main chains are fixed in rocks at each extremity; the western approach is by a gateway formed in the old town wall, and by an embrasured terrace around the basement of one of the towers; the masonry of the supporting pyramids, and also the toll-house is made to correspond with the old castle.

Waterloo Bridge.

Where the Shrewsbury road crosses the Conway river, above Llanrwst, it was necessary to build a new bridge of one arch, 105 feet span; and building stone of proper dimensions and quality not being to be had at any moderate expense, this bridge is built of cast iron. The main ribs consist of the following words in roman capitals:— "This bridge was constructed in the same year the battle of Waterloo was fought;"—and having the national emblems, the rose, thistle, and shamrock, in the angles, it becomes a public and lasting testimonial of the action which so splendidly terminated the war.

THE ROAD BETWEEN LONDON AND NORTH WALES.

Highgate Archway Road.

This road being upon a clay soil with springs of water, originally very imperfectly made, never properly repaired, and at last totally neglected, it became absolutely necessary to thoroughly re-make the whole upon proper principles.

The roadway bottom was, therefore, completely opened, and numerous side and cross drains constructed, so as to carry off the water. Next, in order effectually to prevent the water, or even the damp from the clay, affecting the roadway, a bed of concrete, composed of Parker's cement and washed gravel, six inches in thickness, was laid over it, which, at the same time, formed a substantial bottoming for the road metalling: upon the bed thus prepared, there has been laid a coating of broken Guernsey granite.

There has been a regular footpath formed along each side of the road: the slopes of the deep cutting on each side of the archway, which were cracking and slipping down, have been dressed and sown with grass seeds. The whole remaining now in a perfect state after a trying winter, is an instance that even a seemingly desperate case may, by proper exertions and skill, be effectually remedied; and also proves of what importance it is to have a road very perfectly made at first.—Distance improved, 1 mile 892 yards.

Barnet and South Mims Road.

Between the town of Barnet and the Village of South Mims, an entirely new road has been made, with two bridges and a regular footpath; proper recesses have been made for containing repair stones, clear of the roadway; the toll-house and milestones are of the same plan as those in North Wales.—Length, 3 miles 352 yards.

St. Albans Road.

A new road has been made from the Red Lion Inn, in the town of St. Albans, across the river Vere to Pond Yards, with considerable cuttings and embankings, and a bridge over the river.—Length, 2 miles.

Hockliffe Hills.

An extensive improvement has been made at Hockliffe Hills, consisting wholly of deep cuttings and embankings; the roadway is formed with rubble stone pavement bottoming, covered with broken pebbles.—Length, 1 mile 1,672 yards.

Sandhouse Hills.

This improvement consists wholly of cuttings and embankings; the roadway is constructed as the last mentioned.—Length, 1,320 yards.

Brickhill.

There is a new piece of road at Brickhill, chiefly cutting and embanking; the roadway same as last.—Length, 880 yards.

Fenny Stratford.

At this village, the hollow west of the bridge has been raised by lowering the hill in the street, the roadway has been widened, several houses have been removed, and others underbuilt; fence walls, railings, and stairs have been constructed.—Length, 451 yards.

Old Stratford.

At Old Stratford village the road has been raised, widened, and made with paved bottom, and coated with Mount Sorrel stone.—Length, 370 yards.

Gullet Hills.

This improvement consists of cuttings and embankings; the roadway is constructed with a paved bottom, the workable part covered with broken limestone.—Length, 1,540 yards.

Cuttle Mill.

In this valley there is an embankment 44 feet in height, a cutting 15 feet in depth, a new bridge, and a pavement roadway.—Length, 1,452 yards. There is an additional piece of repaired road:—Length, 400 yards.

Towcester.

Some banking and cutting, and a new pavement roadway.—Length, 247 yards.

Between Towcester and Foster's Booth.

Six hills cut down and hollows filled, and a new pavement roadway made over them.—Length, 1,178 yards.

Stowe Hills.

Over these hills there are several very considerable cuttings and embankings; the roadway is made with a rubble-stone pavement, having a coating of broken Hartshill stone as road metal.—Length, 1 mile 1,120 yards.

Braunston Hill.

This improvement consists chiefly of cutting and embanking; the roadway is constructed, as the last-mentioned, with Hartshill stone.—Length, 1 mile 306 yards.

East of Coventry.

The new road east of Coventry has considerable cuttings and embanking; the roadway is similar to Stowe Hills, with a paved bottom and Hartshill top metal; there is a considerable new bridge, with recesses for stone depôts.—Length, 1 mile 272 yards.

West of Coventry.

The new road between the city of Coventry and the village of Allesley is the same in all respects as the last mentioned, as to roadway, bridges, and depôts; but, in addition, there are two new toll-houses and gates.—Length, 2 miles 240 yards.

Pickford Brook to Meriden.

At Meriden a new road has been made by very considerable rock cutting and embanking; the roadway is paved with hard sandstones, and coated with hard Warwickshire pebbles properly broken. From Meriden Hill to Pickford Brook Hill the road has been put into a proper form, and coated with six inches of broken pebbles. A new road has been made down Pickford Brook Hill, with a paved bottoming, coated with broken pebbles. This road from Meriden to Pickford Brook has scarcely needed any repairs during the last six years.—Length, 2 miles 88 yards.

Wednesbury.

Here a new road has been made across collieries, &c. below the town, in order to avoid a steep hill, and save a considerable distance. The bottom of the roadway is paved and coated with broken stones.—Length, 1 mile 704 yards.

Bilstone.

This new improvement saves passing along a very awkward street, and is only about half the distance; it is constructed in the same manner as the last mentioned.—Length, 1,150 yards.

Wolverhampton.

The Wolverhampton improvement has been carried partly over a level field, and partly through old houses in the town; the roadway bottom is made with large cinders, and coated with Rowley rag and Pouck Hill stone.—Length, 1,199 yards.

Summerhouse Hill.

The first contract proceeded from the public-house downwards to near Bonigale public-house; it is all cutting and embanking; the bottom of the roadway is paved, and the top covered with broken pebbles.—Length, 858 yards.

The second contract consisted in lowering the hill above the public-house, and embanking the hollow east of the hill; the roadway made as the last.—Length, 1,584 yards.

Cosford Brook.

In this formerly steep and dangerous pass there is very considerable cutting in rock, and embanking; a bridge has been widened and raised; the roadway made as at Summerhouse.—Length, 700 yards.

Town of Shiffnal.

This improvement is partly over fields on the west of the town, and partly through some houses adjoining the market-place, and across a brook, where a new bridge has been built; the bottom of the roadway is paved, the top is of broken pebbles.—Length, 456 yards.

Knowles Bank.

Consists wholly of cutting and embanking, and making the roadway same as last.—Length, 957 yards.

Llewellyn.

Here a new road, from the last-mentioned improvement, has been made to Prior's Leigh, in the same way as that at Shiffnal.—Length, 1689 yards.

Prior's Leigh.

At Prior's Leigh there is much cutting and embanking; the roadway has a paved bottom and a coating of broken whinstone; the whole is fenced by stone walls.—Length, 1724 yards.

Ketley Works.

At Ketley Iron-works an improvement has been made, consisting of cuttings, and an embankment 23 feet in height, also an arched roadway under it; the roadway has a paved bottom, with a covering of broken stone; the fences are of cinders from the iron-works.—Length, 816 yards.

Gobowen to Chirk Bridge.

From the village of Gobowen to Chirk Bridge there is an entirely new road; the bottom is paved and covered with a coating of broken stone. There is one bridge over the canal.—Length, 2 miles 1452 yards.

It is fit to observe, before I conclude this description of the roads made in England, that the work has been chiefly confined to cutting and lowering hills, and forming long and high embankments, so that the greater part of the expense has been incurred in moving earth.

HARBOURS.

Holyhead Harbour.

The money granted for this harbour, since it was put under the Holyhead Road Commissioners in 1823, has been expended in giving additional security to the great pier, the lighthouse, and roadway; in deepening the harbour, and laying down moorings for above fifty sail of vessels: this last work relieves the space allotted to the post-office packets. A large anchor has also been laid down at the north pier head, for their security in stormy weather.

The graving-dock, which was in hand in 1823, has been completed; proper gates have been put up, and also a Boulton and Watt's engine; a carpenter's shop, smithy, and storehouse have been built; also a gatehouse and a boundary wall, enclosing the dockyard. A road has been made from the town of Holyhead, communicating with the dock along the south side of the harbour.

About two thirds of the south protecting pier have been built, and backed with rubble stone.

On the north side of the harbour, at the root of the great pier, a lock-house has been built; also a harbour-master's office, with a turret clock-house and clock, and a custom-house.

Lamp-posts and lamps, posts and chains, have been put up, and good foot-paths have been formed along each side of a new-made road between the landing pier and the principal inn; and a coach-house and workshop, for the use of the mail coaches have been built adjoining the inn.

A considerable quantity of rubbish has been removed from the harbour by the diving bell, and the landing place at the pier has been rendered more commodious.

Howth Harbour.

The money granted for Howth Harbour, since 1823, has been applied as follows: —

In renewing the railroad between the harbour and the quarries at Kilrock, providing a quantity of suitable stones, applying a part of them in strengthening the glacis at the back of the eastern pier and that side of the harbour, and having a quantity in readiness in case of injury from storms.

The roadways upon and adjacent to the piers have been put into a perfect state, and the inner edges of the quays have, in part, been secured by posts and chains.

The entrance to the harbour and the packet births have been deepened by taking up 5963 tons of rock by means of diving bells, and 19,967 tons of sand and mud by dredging machines; thereby affording 11 feet of water at low water of ordinary spring tides, where there was only eight feet before.

HOWTH ROAD.

Between Howth harbour and Dublin, a distance of eight miles, the road (formerly very imperfect) has been wholly re-made, and rendered in all respects similar to the Holyhead road; it is now in a perfect state, having a proper cross section, and being smooth and substantial. A considerable sea-wall has been built to protect the road. It is now referred to as a model for other roads in the vicinity of Dublin.

SWILLY ROCKS.

In the Menai Strait, immediately to the westward of the site of the new bridge, the navigation was rendered inconvenient by the strong currents acting upon a parcel of rocks known by the name of the Swillys; and it was stated by nautical men, that, unless these were so far removed as to lessen these currents, the difficulties and risk would be increased by the new bridge.

Considerable exertions were therefore made, whereby not only have the several rocks complained of been sufficiently cut away, but a projecting point of land has been taken off; and, upon the whole, it is acknowledged that the navigation generally is much improved, and that no inconvenience is experienced from the erection of the bridge.

THOMAS TELFORD.

The Committee having called for an account of the several contracts which have been entered into by the Commissioners, find that one hundred and fifty-one contracts have been made for carrying into execution the several works already described. They also find that all the works contracted for have been executed for the stipulated sums, except in one instance, wherein an exceeding took place of 76*l.* 15*s.*

The Committee beg to refer to the evidence of Mr. Telford for an explanation of the mode by which the contracts have been managed, and to the evidence of Mr. Milne for an explanation of the manner in which payments have been made to the contractors.

An account, which is given in the Appendix, contains the salaries and other charges paid under the direction of the Commissioners.

It appears that the total sum paid for works amounts to 697,637*l.* 10*s.* 6*d.*; and that the sum paid in fifteen years for charges of management amounts to 28,460*l.* 4*s.* 1*d.* This charge is something under four and a quarter per cent. on the expenditure. The sum of 4,583*l.* 4*s.* 7*d.* has been paid for parliamentary fees in passing Acts, and for exchequer fees; and 2,821*l.* 8*s.* 5*d.* for solicitors bills for passing Acts of Parliament, and other general business.

The Committee find that the Commissioners, immediately upon the harbours of Holyhead and Howth being placed under their management, in the year 1823, reduced the amount of the salaries to officers 611*l.* 7*s.* 6*d.* a year, and that they have subsequently dispensed with the services of two assistant engineers: the number now employed is, one between London and North Wales, another in North Wales, who has the care of the suspension bridges and the harbour of Holyhead, and superintends all the road business of the Commissioners; and a third, who has the care of the harbour at Howth, and of the road from Howth to Dublin.

In the Appendix an account is given of the additional Tolls levied for the repayment of loans advanced for the improvement of the road between London and North Wales; the amount of the loans advanced by the Commissioners for the loan of exchequer bills on the credit of these tolls is 44,000*l.*, and the amount of the repayments which have been made is 32,781*l.*

Sir Henry Parnell and Mr. Telford have been examined

on the present state of the roads, embankments, bridges, and harbours; and the Committee have to represent, that these works are in a perfect condition, and likely to continue so, in consequence of the complete and durable manner in which they have been constructed.

The Committee beg to refer to the letters, which they have inserted in the Appendix, of several coach proprietors and coachmen, as showing the benefits derived from one of the last pieces of road-work which have been executed by the Commissioners.*

The Committee have to observe, that, although the expenditure on these works, in the course of the last fifteen years, has been considerable, great advantages have been derived to the public from the improved state of the road, and the more rapid and regular communication between England and Ireland; a large saving has been effected by several measures of public economy which have been adopted in consequence of the improvement in the communication between London and Dublin: such, for instance, as the abolition of the separate revenue boards, and the transferring of the chief management of all the revenue affairs of Ireland to London. An annual sum of 12,000*l.* has also been saved, which was expended, before the roads were improved, in maintaining an express establishment for carrying the correspondence of government between London and Dublin; and the postage revenue on letters passing between Dublin and England has considerably increased.

As all the works have been executed by contract and competition, and as it appears that several of the contractors have failed, the Committee consider this a proof that the prices at which the contracts were made could not have been beyond what were fair and sufficient.

* This alludes to the Archway road: a few of the letters referred to are given in a previous chapter.

The Committee, on the whole, feel themselves justified in saying, from their inquiries into the proceedings of the Commissioners, that the works executed by them afford an example of road-making on perfect principles, and with complete success; and, in making this Report to the House, they cannot conclude without stating their high sense of the public and permanent benefit which has resulted from the unexampled exertions of Sir Henry Parnell, in discharging his duties as a commissioner of the Holyhead road, and from the great skill displayed by Mr. Telford in overcoming the seemingly insuperable difficulty of erecting a bridge over the Menai Strait, and also in every other work which he has executed.*

* Since 1830, further improvements have been made by the commissioners on the English portion of the road, viz. at Flampstead Hill, Chalk Hill, High Ash Hill, Denbigh Hall Hill, Geese Bridge Valley, Knightlow, Willenhall, and Allesley Hills, and at Mountford Hill.

APPENDIX, No. V.

THE FOLLOWING ARE THE EXAMINATIONS BEFORE THE COMMITTEE OF THE HOUSE OF COMMONS IN 1836 OF MR. JOHN PROVIS, MR. JOHN M'NIELL, AND DR. LARDNER.

MR. JOHN PROVIS, clerk and surveyor of the Shrewsbury and Holyhead road, called in; and examined.

What situation did you fill before you were appointed to that office?—I was employed upon the improvement of the Holyhead road, and also in proving the iron-work of the Menai Bridge, under Mr. Telford.

For how many years were you under Mr. Telford?—Fourteen years.

You have received a regular education as a civil engineer?—Yes.

I believe you inspected several of the large contracts, under Mr. Telford, for making the Holyhead road?—Several of them.

Will you describe to the Committee the process of making a road under a contract, namely, the sort of contract and specification, drawings and inspecting, when a new piece of road is to be made; what course did Mr. Telford take?—After the survey had been made and decided upon, the first operation was that of forming the road; after that, of setting on it a pavement of sixteen feet in width, and averaging six inches in thickness, that is seven inches in the middle and five inches on the sides; and upon that six inches in depth of broken stones, with a slight covering of fine gravel, to prevent the horses feet being injured.

What was the width of that road?—Thirty-two feet.

What were the fences?—Stone walls, four feet six inches in height, including a coping on the top.

Was this mode of making the road described in the contract?—It was.

How?—It was described in the contract by drawings and a specification pointing to each particular, and the manner in which each separate work was to be performed.

And how was the work superintended, so as to secure the performance of the contract strictly according to the agreement?—As soon as the contract was made there was an inspector appointed to superintend the work.

And what did he do?—He had to examine it daily, or as often as it was necessary; and then, if he found any thing wrong, it was his duty to point it out to the general surveyor or engineer.

Did you act as inspector in any contract?—Yes, I did, to several contracts.

So that you were always enabled to see whether the contractor had faithfully performed what he had agreed to do?—I did; I saw it properly formed before the pavement was put on, and I also saw the paving before the broken stones were laid on.

Was there any particular regularity about the pavement, about the size of the stones?—There was; that none of them should exceed four inches in its upper surface.

In breadth?—In breadth.

What length?—Generally about eight or nine inches in length.

How were they set?—All set by hand, with the broadest edge downwards.

What class of workmen were employed in setting the pavement?—Labourers who, from long practice, had been accustomed to it.

The upper part of the pavement, how was that managed? —The paving was laid by hand; the stones were laid as close as they could in parallel layers, and after that the interstices were filled up by small chippings, forced in by a hammer.

So that the whole mass became a solid compact body?—The whole mass became a solid body.

No stone could rise easily up?—No, it was almost impossible.

What sort of stones did he use?—A variety; it depended on the nature of the quarries.

Was it necessary to have it hard?—Wherever we could get it; it was not necessary to have the lower stratum so hard as the upper.

They must be so hard as not to decompose?—Yes, they must.

What size did you break the stones?—So that the greatest dimensions of the stones should pass through a ring two inches and a half in diameter.

Was this the practice on the Holyhead road?—It was.

Do you know when Mr. Telford first began in the Holyhead road?—It was in 1815, I believe.

What is the distance of what may be called the new road?—Eighty-seven miles.

From what point?—From Holyhead to Gobowen.

Is all the road either entirely newly made, or so changed as to be fairly called a new road?—The whole of it.

Is all of it paved?—The whole, with scarcely any exception.

Some part of the old road was widened only; were they new paved?—Almost all of them, I believe, through North Wales.

What is the breadth of the road?—Varying from twenty-two feet, where we have precipices or difficulties to overcome, to thirty-two feet.

The ordinary breadth is thirty-two, accurately?—Yes.

What is the convexity?—Eight inches.

Was that quite accurate so that it would prove to be eight inches by using a level when the road was finished?

—Yes, we have templets of the shape of the road, but it is accurately ascertained as the materials are put down, the road being first levelled, and then measuring the depth of the pavement at the ends and the centre.

Then the road has not the appearance of a common road, sometimes high on one side and sometimes low on the other?—No.

It is altogether a construction, as it were, made on fixed rules, and according to the particular measures?—It is; and wherever in one part it is narrower than others, it is always brought so gradually, so as to produce no abruptness in the appearance of the fences.

Then when you do reduce the breadth to twenty-two or any smaller breadth than thirty-two feet, you continue that until you pass the ground which is difficult?—We do.

Is there a uniformity?—Always.

The fences, you say, are walls?—Stone walls.

Your convexity is not the same when it is narrowed, only proportional?—Proportionally so.

Then your templet would be no use there, because it would be too high—We principally use the spirit level.

The road being through a mountainous country, crosses a great many small rivulets and streams?—Yes.

In regard to drainage is it perfectly well drained?—Perfectly well.

Do you find that is the case in very wet seasons, and upon sudden heavy falls of rain?—I have never yet found any difficulty in getting the water off; it is never any impediment to the travelling.

Does it ever injure the surface of the road, so as to carry away parts of it?—Not of any consequence.

Are these pipes and small culverts and bridges built in a very strong manner?—Yes, they are.

Were there particular pains taken to secure the foundation?—A great deal of care; in many instances they are upon a rock.

Were they paved under, so as to prevent the water cutting under the side walls?—All, except where upon a rock.

A regular rule?—A regular rule.

Now, in severe winters, after very long frosts and sudden thaws, does the surface break up?—I never saw an instance of it in any part that had been paved.

That is in such winters of such severity that in ordinary cases the roads generally break up more or less throughout the whole country?—Where other roads under my charge have broken up.

Then, in summer, does the surface loosen and show weakness?—No symptoms of weakness; occasionally after a long continued dry season it would show symptoms of looseness.

The stone will not rise so as to cover the surface?—No.

Did you ever find the pavement stone rise up?—Never.

Does the pavement serve in any degree as a drain to the road?—There is no doubt of it.

Have you a communication between the pavement and the drains of the road?—Generally speaking we have.

In what cases have you not?—We have not a continued communication, but we have loose materials of a similar sort to communicate with.

You have a sufficient number to carry off any water that collects on the pavement?—Yes.

As there were stone walls there could be no necessity for ditches?—No; there are ditches in some places, but they are of rare occurrence; we seldom require them, because the field-drains being sufficiently near, we try to communicate with them wherever we can; we always drain them to the lower side; the greater portion of the road is on hanging ground.

The water that collects at the upper part of the road is carried to cross drains?—Carried across by a drain below the road, and in most places, where we have hanging ground, the road is supported by a breast-wall, and the water merely passes over this breast-wall into the field.

How do you prevent the water falling from the high ground into the road?—By leading it in channels into the cross drains.

There are drains in the upper part of the road that carry any water that falls from the upper land into the cross-drain, and keeps the water from the surface?—Yes.

In forming the road, in point of fact, is it not the case that the road is elevated, the surface of it, so as to be above, as it were, the immediate ground that touches it?—It is the case; we always prefer raising the road to depressing it.

That is, you always form the road so as to elevate the sub-soil of it on which the materials are put above the level of the adjoining field?—As much as we possibly can.

Which in itself is a protection from moisture and wet getting upon it?—Yes.

Is there a wall on each side of the road?—There is, for nearly the whole way, except where we cut through rock.

Is that wall on the hill side somewhat bastion fashion, a little on a slope, so as to act as a buttress?—Not a buttress; but it is wider at the foundation than at the top.

The form of the road, the convexity, in fact, produces a side channel towards the higher side in which all rain-water falls into the cross drain?—All surface water runs in the channel on each side, close to the wall, until it gets to one of the cross drains.

So that the road is so formed when crossing a sloping ground that the water cannot rise from the side channel next the hill, to the middle of the road?—It cannot.

Through what does the water escape when it gets to the position of one of these drains; how does it find its way; what orifice is there at the surface?—We in general make them communicate with the field drain; and we have an opportunity of joining them to the field drains at very short distances.

How are the inlets to those drains protected from the silt choking them up?—Generally by masonry, or paving the mouths with large stones.

Is there much silt on the road?—In parts.

What sort of materials is the upper coating of; broken stone?—A good deal depends on the rock through which we pass; we always select the best.

Is it generally all through very hard?—It is of variable quality in that respect.

Have you much scraping?—Very little in some parts, in some parts none.

What makes the difference?—The hardness of the stone, and the difference of the traffic upon it.

Would you say the hardness of the stone made a great difference in the quantity of scraping?—No doubt.

Is not some part of the road so hard that it never requires scraping?—Yes.

And is not the stone much harder than elsewhere?—The stone is very good in that place, but that is not the only advantage; it has a good exposure to the wind.

What part of the road is it?—Lake Ogwen.

Where have you the most scraping to do?—Between Chirk and Gobowen.

What is the material used there?—Chiefly limestone.

Is there a heavy carriage?—Very.

What sort?—Principally of coal and lime.

What is the quality of the stone where the scraping is the least?—Basaltic.

What is the road across Anglesey; is that hard?—Yes.

Is the material hard there?—The best I have got on the road.

What is it between Corwen and Llangollen?—Principally schistus.

Cannot you get harder stone?—Not without bringing it from a considerable distance.

Is there much traffic there?—A good deal; of coal particularly.

Have you tried the newly invented scraping machine?—I have, for a short time.

How do you find it answer?—Remarkably well, as far as I could judge. I do not like to give a decisive opinion on it till I have tried it longer.

What is it, as far as you can form your opinion, in point of expense, with reference to the usual mode of scraping?—I think we shall be able to scrape at half the expense.

What does the machine cost?—Six guineas and a half.

One man will do the work of two?—Yes.

What is the distance of the road from Shrewsbury to Holyhead, which is under the management of the commissioners?—There are 106 miles, besides three miles of branch road, that is 109 altogether.

This road was formerly, before 1819, under the management of separate trusts?—It was under the management of seven separate trusts.

By what act was the consolidation effected?—The 59 Geo. 3. c. 30.

What was the plan of consolidation with regard to the trusts; how was that arranged?—By agreement between the commissioners and the various trustees.

The agreement had reference to the debts?—Yes.

And how were the debts settled?—A portion of the debts belonging to each of these seven trusts was placed upon the Shrewsbury and Holyhead road.

Those trustees had other roads under their management, besides the main line of road?—Yes.

What is the debt upon the road?—It is about 17,000*l.*

Has the interest been regularly paid since 1819?—Regularly.

Have the rates of toll on this road been much reduced

of late?—Very considerably; perhaps I ought not to say of late, but soon after the act of 1819 was passed, they were 1*s.* 3*d.* for each post-horse.

At how many gates?—At seven gates, between Bangor and Llangollen; from that they were reduced to 1*s.* a horse, and from that to 9*d.*, which they are at present.

And have the rates of toll upon stage-coaches been reduced?—Very materially.

Distinct from post-horses?—Distinct from post-horses.

As to the annual amount of tolls collected, what comparison does the state of the income this year and the last bear with reference to past years?—The lettings of tolls this year have exceeded any thing I recollect since I have been on the road.

How much are they beyond what they were last year? —£670.

Have the tolls of the Menai Bridge become increased this year?—Very much.

How much?—£300.

What is the annual amount?—The present letting is 1,025*l.*

From the time that you have been employed on roads, you are acquainted with roads that are not made with the regular pavement, are you?—Yes, I am.

What, in your opinion, from your practice, would you say was the difference, with regard to the object of having a good road, between the system of the road made with a foundation of pavement, and one made by broken stones on a sub-soil, as regards the object of a good road generally?—Decidedly, I prefer a pavement.

With regard to expense?—With regard to expense, I have no doubt it would be cheaper.

Do you find the surface materials wear longer when they are laid upon a paved bottom than when they are laid on a sub-soil?—I have scarcely had an opportunity of judging as to the comparative difference.

Upon your road you have no means of making that comparison?—I have not.

I think you stated that a paved road is cheaper, and wears longer, and yet I think in a former part of your evidence, about a quarter of an hour ago, you said the breadth was thirty-two feet, and you have given to the middle sixteen feet. If it is harder, more durable, and cheaper, why do you not make the whole road so?—I believe it is stronger and cheaper.

Then why not make the whole road so?—It is cheaper than breaking an equal quantity of stone. Say I make my road with materials a foot in depth; if I broke that foot in depth I could pave it, and metal it with six inches at the top cheaper by one third than I could break the whole of it.

With regard to paving sixteen feet in breadth, have not the roads recently made by Mr. Telford in England where the traffic is considerable been paved from side to side, that is thirty feet wide?—In two instances, where I have made roads under Mr. Telford, they have been paved across from side to side.

Suppose a road paved across from side to side, and coated with broken stones from side to side, would you say that such a road might be made cheaper with a foundation of pavement than with broken stone laid upon the surface, supposing the thickness of the road of broken stone to be about ten inches?—I am convinced it would be cheaper.

Have you made a calculation upon that point?—I have; the average price paid for good stone in the county of Anglesey is:

	s.	d.
Quarrying, per cubic yard - -	0	8
Breaking, per ditto - - -	1	10
Total for quarrying and breaking	2	6

The expense of paving the middle sixteen feet in the usual way, with six inches of broken stone on it, will be, per yard lineal, 3*s.* 0½*d.*

The same width and depth of materials, but the lower six inches broken instead of being laid as a pavement, will be, per yard lineal, 4*s.* 4½*d.*

The same width without pavement, but having ten inches in depth of broken stone, will be, per lineal yard, 3s. 8¼*d.*

In the latter case there will be a saving of one third of a cube yard of stone, which, supposing the cartage to be worth 9*d.* per yard cube, will reduce the cost 2*d.* per yard lineal.

What is the composition of the other part which is not paved?—The upper part is of broken stone.

To what depth?—Six inches.

In your calculations do you confine yourself to the sixteen feet?—I do.

It is nothing to say there is the side between the sixteen feet and the fences?—That is of no consequence; the same proportion will go through the whole.

That is, a road sixteen feet wide made with a pavement, will be cheaper than a road sixteen feet wide made with broken stones?—Yes.

If you double the depth it will of course be still cheaper? —Yes.

If you were to make a road now, would you make it sixteen feet or thirty-two feet?—I would make it thirty-two feet if I had plenty of materials.

Would that not depend on the traffic?—It would a good deal.

In the greater part of Wales the breadth of sixteen feet will be sufficient for the traffic when the road was first made?—It was.

Have you not by degrees strengthened the sides according as you have had funds?—Nearly the whole of them.

That is, you have laid on coats of stone, and brought them to be at present in a perfectly solid state?—Yes.

As this happens to have been an old system (for I see here in 1819 the following question put; "Is not a road constructed with a roadway of sixteen feet breadth of solid materials, and with six feet on each side of that with slighter materials, a sufficient road for the general purposes of country travelling?" this being before the Committee in 1819), if the old system works so well, how comes a new system to be so generally adopted of laying the stones on the road, such as we see in Oxford-street, without any sub-pavement?—That I am not able to account for.

Would not, in your opinion, Oxford-street be a better street, and the materials wear less quickly out, if there was a pavement laid down before the broken stones were laid on?—Yes; I noticed, going down Whitehall, that that road is irregularly sinking. I can only account for that from the weakness of its foundation.

How can you account for the great quantity of dirt that you see on the surface of the London streets, except by the working up of the dirt below?—I cannot account for it, except from the constant traffic over it.

When small stones are laid upon a soil, is it not the case that they sink into the soil, and the soil rises up between them?—It naturally must, if they are thin.

When you talk of a road being sixteen feet wide in the middle upon a total breadth of thirty-two feet, how do you support the pavement on the sides; how do you keep that pavement in its place?—The materials are sufficient; all the stones are placed lengthways; their abutment is chiefly against each other.

What is the abutment of each extremity?—It does not

require an abutment; it would support itself; we place each stone with the largest end downwards.

Does the road want an abutment on each side?—I do not consider that it does.

Do you consider a road to be in the nature of an arch? —I do not.

Why is it that the stones at the extremity of each of these parallel lines do not give way in case of any regular pressure upon them?—If the pressure was lateral, they would give way; but the pressure is vertical.

The Holyhead Road, I believe, was made by grants of public money from Parliament?—It was.

Were those grants sufficiently large to admit of such an extent of road being made, so as, in the first instance, to make it all a pavement from side to side?—No.

Was that the reason Mr. Telford proceeded on the plan of paving only sixteen feet in the middle, in the first instance?—That is the only reason I ever heard assigned.

As soon as the different parts of the road have been finished, according to the plan on which they were originally constructed, have not the side parts between the fences and the paved part been brought into a perfectly solid state by continually laying on fresh coatings of broken stones?—It has.

So that the whole road now from side to side is a perfectly solid road?—Yes, with very trifling exceptions.

Are the side parts of the Holyhead Road so hard as what may be considered a good turnpike road?—No, I do not think they are; they are less worn; there is less traffic on the sides than in the centre.

Originally the side parts were made of six inches of broken stone or rubbish?—Yes, generally.

And you have been continually laying fresh broken stones upon it?—Where we found places show symptoms of being soft, we cut out the clay, or whatever it might be, and substituted good broken stones.

How deep would you put layers of broken stones?—Generally about three inches in depth.

Have you laid broken stones, so as to keep the surface perfectly hard and free from dirt?—Yes, we have.

You do not find it work through?—Occasionally we have.

Do you find, practically, that almost all the traffic, almost every carriage that runs on the road, runs upon the sixteen feet in the middle?—It does; wherever I have seen an instance of cutting into the sides, it has been from putting fresh materials in the centre, that they go on the side to avoid it.

In point of fact, the centre part of sixteen feet answers the purpose of affording to carriages the advantages of a solid road and easy draft to the horses?—It does.

As to the elasticity of the road, do you consider there should be any elasticity?—Decidedly not.

What do you define to be an elastic road?—A road that will give way under any weight that is passing over it.

You would not define any of those roads which do not yield to be elastic roads?—No.

You would not define Oxford Street to be an elastic road?—I have no means of ascertaining that.

What would you consider your own road?—Certainly not elastic.

Did not Mr. Telford employ the plan of paving in order to destroy all elasticity as much as possible?—Yes.

I see Mr. M'Adam says, that "wherever a road is on a hard foundation, the road wears much sooner than placed on a soft substance?"—I never found an instance of that.

You do not think materials are more likely to wear on a hard substance than a soft?—They do not, if there is a good body of them.

Supposing you wanted to crush a lump of sugar, would you not put it on a marble slab?—Yes; but if I did not want to crush it, I should put other lumps by the side of it.

Are you not of opinion that if you have a quantity of broken lumps of sugar, and you put them on a piece of dough, they would be much less likely to crush than if you put them on stone?—Yes; they would sink into the dough.

How long have you been a surveyor on the Holyhead Road?—Ten years.

Have you had much experience of the crushing of materials on roads so as to be able to state whether they are liable to be crushed on a paved bottom?—We keep it too well coated to give it an opportunity.

From your experience there is not any foundation whatever for the common notion that stones are crushed and destroyed when there is a pavement between them and the soil?—There is not when there is a good body of materials on the pavement.

What number of cubic yards per mile do you use upon the Anglesey Road for the common repairs of it?—Ninety-six yards were last year put on it.

And that kept the road up?—Yes.

There is a good deal of travelling over this?—There is a good deal, but I can hardly call it very considerable.

How many coaches pass it every day?—Four each way daily; two mails and two coaches.

Nothing else?—No more.

Are there not lime carriages?—Yes, lime and coal.

What is the quantity of materials you put on per mile between Bangor and Cernioge?—Seventy-three yards per mile.

That keeps the road in perfect order?—It is in beautiful order.

Where you have more traffic, namely, from that point to Chirk, what is the quantity of materials you put on per mile?—One hundred and sixty-two yards per mile.

It is with those quantities that you keep the Holyhead Road in that order which every body admits, generally speaking, to be very good?—It is.

You do not find, in point of fact, that sort of crushing takes place which destroys the materials very rapidly?—No, I think quite the reverse.

There is no regular laying on of this 162 yards per mile; they are all laid on according to the exigencies of the place?—Just as they are wanted.

No regular three or four inch layers?—No.

Is the road kept up to a proper degree of thickness?—I have every reason to believe it is.

There is no wearing on the road?—No.

Have you had any opportunity of making a comparison of materials?—No.

What is the effect of an elastic road on the draught; do you consider it increases the labour?—Yes.

You stated, if I understand you right, that the putting hard materials upon a hard road did not occasion the crushing of those materials; I ask you, from your own observations in this town, if you were to put a bushel of those pieces of broken granite in the middle of Oxford Street, would they or would they not be crushed? Let us assume a bushel of broken granite to be laid down without any picking it, that it may amalgamate with the other, put that bushel of broken granite upon a granite pavement, and let the traffic go on it, would not, in the course of half an hour, those stones be crushed?—A great many of them would, no doubt,

Would you, as an engineer, do such a thing as lay a bushel or two of stones in the streets?—Decidedly not.

Why would you not?—If I understand your question, the only reason that I could put that quantity on was, that there was some little part that was out of shape.

If you had to repair the streets of London, you would put on a sufficiently thick coat to prevent the crush that would take place, and not put it a single coat thick?—Yes.

Does it not strike you, that before you put this bushel of broken stones or any quantity of broken granite on the road, that you would pick up the road, in order to present a surface for the stones that they might sink in?—I always put it on when the road is soft.

You do not mean otherwise soft than just during the wet season?—I always put them on in the wet season.

Why do you put them on in the wet season?—Because I always find the stones will bind much better.

Do you not mean to say by binding, that there is a softness of surface, occasioned by the moistness, so that they sink in more easily by pressure, and avoiding the crush?—There is a greater degree of softness in wet weather than dry weather; they get better hold of the road in the wet season than in the dry.

You have observed that in London the practice is, before they lay on a coating of broken stones, to pick up an inch or two?—Yes.

Is not that an expensive process?—I conceive very much so.

It is a difficult process, is it not: a man cannot get on so fast?—Yes.

Therefore a considerable expense is incurred by the operation of picking?—Yes.

Is there not also a great waste of materials by striking the surface with the pickaxe?—I should certainly suppose so.

Then in point of fact the operation is of a very expensive character?—Yes.

In your practice, you say you do not have recourse to it; that you lay on the materials, when the road requires repair, in the wet season; and do you find it bind perfectly well without?—Yes, I do.

In point of fact from your experience in repairing this long line of road, there is no occasion for this expensive operation?—I never find that there is.

In some parts of the road I believe there is considerable traffic, which would show if there was any reason for the course being adopted that is adopted in London, in order to prevent what is conceived to be the crushing of materials; there is as much traffic in parts of the road where the coal carriages go, as in some parts of the streets of London?—I should think so, very nearly.

When you spoke of putting on the materials in wet weather, is it because the surface is soft at that time?—It is not; it is because the materials are less firmly bound together in wet weather, and more easily allow the new coating to sink into the road than in dry weather.

It is not soft?—It is not in a state of mud.

It is so far less hard than in dry weather, as is natural from rain falling upon it?—Yes.

What is your object then, in putting this material on it in wet weather, if it is not a softer substratum?—Because it is easier to get it to set amongst the other stones.

It is necessary to have a thicker coat of materials?—I never adopt that proceeding of picking up the stones.

What is your rule with regard to the time of year in which you make your repairs by laying on new coatings?—I always begin in September, that I may have the advantage of all the wet weather to consolidate and to set the materials; I never, when I can avoid it, put a particle of stone on the road in the summer.

You put it on in that time of the year that the substratum may be soft, and by that means these stones may unite with the crust; assuming that you have ever such wet weather, and you were to lay stones upon this table, and all the rain of heaven came down upon these nine inches of granite stone or basaltic stone; if nine inches were put upon this table, and it was exposed to the weather for sixteen years, would they ever unite together?—I never put stones on a pavement such as this table, they would keep moving round.

Do you mean that hard granite on a substance like this would bind together?—By their own pressure they would; all stones bind by crushing together.

What is your practice, supposing you have to make a road over a rock?—We usually level it as well as we can; we never can get it to a perfectly level surface; we level it first by quarrying and getting it into the general shape we want for the road; upon that we occasionally have paved, but generally the practice is to break the stones and then to lay a body of eight or nine inches of them on the top of it. Where we do not pave we always give a greater thickness of broken stone upon it.

You do pave over the rocks?—In many instances we do.

If you have a perfectly flat hard surface of rock, you would pave on that?—Yes.

Why?—Because the flat smooth stone would allow of motion upon it; the generality of rocks we blast have a good rough surface.

Have you formed any opinion as to the different force required, or the labour of horses in drawing carriages over roads made with pavements and made without them?—I should think it would be less, but that is merely matter of opinion.

Have you ever had any opportunity of seeing any statement upon that subject?—I have.

Where have you seen it?—In some of the Holyhead Road Reports. I think in the last made by Mr. Telford, there is a statement of some experiment made by a machine of Mr. Macneill's invention.

Do those experiments show a difference of draught?—As far as I can recollect, they show a difference in favour of the paved system.

Are you aware that Mr. M'Adam said, that in a road near Bristol the expenditure was seven on a hard surface in comparison to five on a soft, the soft being over a bog? —I never heard that.

Your experience does not accord with that view of the case?—No.

In proportion that materials wear longer by being on bog, would that occasion such a degree of elasticity as would add greatly to the labour of the horses in drawing a carriage over a road of that kind?—Yes.

Is not the proper object of a good road to reduce as much as possible the labour of horses upon it?—Yes.

You have stated that the draught is harder for horses on an elastic road?—Yes.

In what way?—Because the weight to be moved is constantly in a hole.

What do you define an elastic road to be?—An elastic road I consider to be that which will give way by pressure, and rise again to its original position when the pressure is removed.

Do you consider the road in Oxford Street to be an elastic road?—I do not know whether I could call it elastic, I know so little of Oxford Street.

Generally speaking, you define an elastic road to be one that yields to pressure. Do you consider then that such a road as you have called a pavement road, but which in fact ought to be called a laid road, do you consider these laid roads, to which you have been alluding, are or are not elastic?—I do not consider them elastic; at least they must be possessed of very much less elasticity than the broken-stone road.

Have you at all considered the eligibility of consolidating the trusts and the funds of trustees of roads?—I have considered it, though perhaps not very maturely.

Do you consider it advisable?—I do.

I think your road is 100 miles in length?—Yes.

Do you think that in consolidation it is advisable to consolidate one great length of road, or roads round a centre?—I should say one great leading road.

Would you give your reasons for that?—Because, generally speaking, if you take some one leading road, there is a greater quantity of traffic, and it requires a different mode of management to side roads; there would be less material required, and less care would do for a side road than would be necessary for an extensive road, such as the Holyhead road.

Supposing the present Holyhead road be left as it is, what species of consolidation do you think might be applied to the other roads that branch out of it?—I do not think I am capable of answering that question.

You are not aware that there are any roads which could be advantageously united to the roads you are at present employed upon?—I am not.

You have not turned your attention to the subject of substituting some other mode of revenue than tolls?—It is a subject that I have sometimes considered, and I think no other mode could be devised better than the present.

Have you thought anything as to the proposition of employing persons engaged on the roads in the police?—I should not like my men to be turned into constables.

Why not?—I should think they could not do their duty in both capacities; they would be taken away for various purposes unconnected with their duty to the roads.

Would it not produce a great derangement in the management of the work?—I think it would.

Could you obtain as much work from your labourers if they were allowed to act as police as you now do as labourers only?—I think they would be often taken away from their work to act as police officers.

Could you exercise so much control over them as to secure proper attention to the work which they have to perform?—I should very much doubt that.

Workmen come at a certain hour in the morning and leave at a certain hour in the evening; could you, as a responsible servant to the commissioners, get as much labour done by your labourers as you do now, if those labourers were also employed in the police?—I should be afraid not.

Then, in point of fact, this plan of making them a police would be an additional expense to the commissioners?—Yes, I conceive it would.

And, at the same time, render the labourer less efficient?—Less efficient.

You mean to say, that if they were taken from their work they could not do so much work for you?—They could not.

Have you at all turned your attention to the question whether or not it would be advantageous to the public to have such a system of road constables by paying a little more additional to those individuals, and to take them off their work if it was required, which might not be often? Have you turned it at all in your mind whether such a system might be adopted with benefit to the country?—I feel convinced it would be of no use on the road on which I am engaged, at all events.

Are the circumstances of the country between Chirk and Holyhead such as to require any scheme of this sort of having a road police?—Decidedly not.

Would they have any thing to do?—Nothing.

Then how would they be taken away from the road?—They might occasionally, but it would be trifling; I do not think I should be so exclusively their master as I am now.

Why, you admit it would be an additional expense, therefore it would be an additional benefit to these men; you then say, that if they were taken off their work they would do less work; you say afterwards, they would have nothing to do as policemen, and therefore the natural result is, they would not have to be taken off their work;

and then you admit they would have a little more pay, and you say that they would be less under your control?—I was wrong in saying they would have nothing to do; I think they would have very little to do.

In point of fact, the relative situation that you would stand in with regard to them, as the employer of them, in the name of the commissioners, would be totally changed; that is, you would not have the same facility of selecting your men, or of removing and dismissing them?—I should not. And there is another point; most of my men work by contract; you could not take them away from their contract work, and make them act as constables.

How many labourers are generally employed on the road?—About 117.

Are they principally employed by contract?—I should say two thirds, at least.

Through the year?—Through the year.

In what way?—Breaking stones by contract; scraping the roads by contract.

Then, in point of fact, you do find this to be an economical and efficient method of carrying on the business of the road?—I do.

More than by employing day labourers?—Yes.

Introducing this system of police would be a derangement of this plan of contract which you have found so effectual?—Yes.

How do you make that out? If your men work by contract, and you pay them so much for so much broken stones, if they are taken off that work they do less by contract, and you pay them less; how is it possible you can make out that you lose by it, or that it would be inconvenient?—Because the men would give up their contracts; they would not be annoyed in this manner; it would entirely derange our system; somebody must pay them for lost time.

You state that these men are men who work by contract, therefore you only pay them in proportion to what they do; then you turn round and say, that those men, if employed elsewhere, would be dissatisfied, because they would not be able to fulfil their contracts; but would they not get remuneration? If a man gets 3*s.* a week more for being a police constable, and gets 1*s.* less a week by contract, why should you be dissatisfied, or why should he?—Because it would take that 1*s.* away from the trust fund; we should have to pay that man for acting as constable, and while he was acting as constable he would be neglecting our work; it would be giving each man two employers.

Suppose your men were liable to be called away for any other duty, could you undertake to say you could discharge your orders as punctually and as effectually?—I could not.

Do you mean to say, if any of your men got drunk, and did not fulfil their contract, that would put you or the road to any inconvenience?—It certainly would, though we do not pay them for drunkenness.

If any of your men, in consequence of getting drunk, could not fulfil their contract on that day, do you mean to say that a loss of that man's contract for a day would be an inconvenience to your trust?—Not that individual man, because I could get another man to do his work.

Then supposing this man was called on to act as a police constable, and did not do his work for one day, how could it be any inconvenience to you?—I conceive that I should have that man's wages to pay.

But you said just now you paid by contract?—If I do not pay him, who does? Somebody must pay him.

We are speaking of this man who does your work by contract; if he does not fulfil his contract you do not pay him?—No.

Then how can you say you have got to pay him if he is off the road?—If we did not pay him it would make no difference.

Suppose a labourer was appointed to act as constable, and that in consequence the power of dismissing him was taken away when you thought proper, would not that be attended with very great inconvenience?—Decidedly.

Could you manage your labourers if you had not the power of instantly dismissing any of them?—No.

Then any plan of interference, by appointing labourers to act as policemen, must necessarily lead to that inconvenient consequence?—If it prevented me having the power of dismissing them, it would.

John M'Neill, Esq., Civil Engineer, (employed on the Holyhead Road under the Commissioners of Woods and Forests,) called in and examined.

Have you made improvements in your instrument for trying the draughts of carriages on roads?—Yes; I have one now under construction that will be very superior to those I first constructed, as that required a professional man to work it, and take down the results. In the present one, if you send it along the road by a common servant, you can get a complete and exact state of the road; as to its resistance to draught, and the power required to draw a carriage over it, and whereabouts that power is exerted, where it is out of repair, and where it is in a good state; it will mark the precise situation of that part of the road.

How is this information given?—By a line drawn on a sheet of paper by a pencil.

As the carriage moves along?—Yes.

Are you still satisfied of the principle of your machine

being a correct one, as to ascertaining the draughts of carriages?—I am quite certain of its being perfectly correct, for we have tried it in some cases, and compared it with a weight hanging over a pulley, and the results were the same. It is only in certain cases where a weight over a pulley could be applied; it could not be done practically over a road of any length.

Has anything occurred as to the soundness of your principle being controverted by other engineers?—No.

Is it generally adopted by them as a correct machine? —Yes, and referred to in very many cases. In France there has been a petition to the Chamber of Deputies, founded on my experiments, relative to the mode of ascertaining the draught of carriages, and the saving by using springs.

You were the author of that algebraical calculation delivered in the Lords?—Yes.

Does it correspond with the results made by the machine?—It was founded on experiments made by the machine; it was a formula that would give the power required to draw a carriage over a road in a section of that road, from data determined by experiments made by the machine.

Is your machine calculated to give the draught on setting the body in motion, or when it is in motion?—Both.

Then it would appear that your former calculations, as to different effects of different roads on the draughts of carriages, are correct?—Yes, quite correct; and they have been confirmed by very many experiments I have made since I was examined before the Committee of the House of Lords.

Then, in fact, that is the general conclusion, that a road is good for its object, namely, of diminishing the draught of a carriage, in the proportion that it is hard and smooth?—The great advantages of the roads appearing by the machine is certainly in proportion to their solidity and their strength, and their want of yielding. If it could be a perfectly solid mass of stone or metal, the least resistance would be presented; that is shown both on stone tramways and on metal tramways, and metal rails. There are some metal tramways laid in Glasgow on rather a steep hill, and it is not at all unusual for a horse to take from two to three tons; that arises merely from the saving in the resistance of the surface, friction being lessened.

That is from the smoothness of the surface?—Yes, from the smoothness and hardness.

So that if clean material of any road nine inches thick were properly beat down, that will not yield?—Nine inches will yield very much.

What, on an old road of nine inches thick?—Yes, with heavy waggons. One of the great advantages arising from Mr. Telford's system of forming roads by large stone pavements, is from the fact that one point is distributed,—that the pressure of the wheels is distributed over a large space. The wheels of the carriage rest on, say, two inches of surface, but that is carried to a large pitching stone below, which rests on the soil, and the weight is distributed over a large surface at the bottom; that is to say, over a surface a foot or nine inches long, and six or eight inches wide; it is lessened very much indeed on the surface that bears on the earth.

You speak of a heavy waggon: is there a distinction between the crushing of the materials and the vibration?—If it is very smooth there is not much crushing. I should perhaps say, where broken granite is placed on the London streets, there is little crushing on the surface, but a good deal below. In some experiments I made some time ago, I found that a great portion of the wear took place near the bottom; the stones there got round after a very short time.

How do you explain that; they would sink into a soft substance, would they not?—Yes; they were all jumbled together; the lower part of the lower stratum was on clay.

Would not the pressure from above, instead of wearing out those angles, only press it down?—No; before it became a solid mass at all, the wheels worked through and the stones were kept in motion, and rubbing against each other from top to bottom; a portion was put down with a very solid foundation, and the same quantity was put over it. I took up a portion of both roads, and we found where the road was about fifteen or sixteen inches thick, and the stones six inches thick over the bottom pavement, they were quite square and as perfect as when put on. In the other case it was not so.

I am well aware that you are a good mathematician; how is it possible that the wear can be greater on that elastic road; for we all know the first principle of mechanics is, that where two substances meet and one yields gently to the other, the friction, or the wear and tear, is not so great?—That is so on the surface and at the point of contact, but where there is a yielding and elasticity in the road there must be motion among the particles with which the road is formed, and this motion produces wear.

But it may be so imperceptible as to be almost evanescent?—Yes, but when it is so there is no motion among the particles.

Do you mean this motion of the stones on the surface continues wearing off the angles or edges, and making them into round pebbles?—Yes, that is what takes place on a weak road. There are two things to be considered, when a road is newly made; there is very great wear indeed, in the first instance, if there be not a pitched foundation; that is different from the wear that takes place when it is nearly consolidated; to bring it into a solid state the wear is great; if there are four inches of broken stone on the top of a pitched foundation, you may get that road in a perfectly hard and consolidated state, by the ordinary travelling over a turnpike road, in about three months; but if you put on the same thickness of stone without a foundation, you will not get it in a perfect state in three or four times as long.

Do you give that opinion from experience?—It is quite certain.

What instance can you refer to in support of the statement of the quickness with which a road is consolidated, where the upper surface of stone is laid on a paved bottom?—The last example I have had is that one near Shrewsbury, between Wellington and Shrewsbury, a road made there very lately. It was opened to the public in the latter end of December, and in the early part of May it was quite as solid and perfect as if it had been in existence for fifty years.

What was the material?—It was a sort of porphyry; it is a hard compact sand-stone, with no lime in it.

Was this made with a paved foundation?—Yes.

If I understand you right, you put such a thickness of broken stone over the pitched bottom they consolidate into one solid mass quicker than if there was no pitched bottom; how do you explain that?—In order to put in stone at all that will bear carriages in the first instance, you must have a certain depth of them. You cannot put in four inches; if you put in four inches the wheels will sink through and the clay will rise; it will be all mud and stone together. Then you must put on three or four inches more and so on, until a sufficient stratum is formed. Until this takes place there is great waste of stone, and you have not got a solid road after all. A great portion of that stone is worn away in the mean time and mixed with mud, liable to get soft with every shower of rain. When you put in nine inches thick of a large stone pavement, the mud cannot get up; the four inches of the small stone put over cannot sink. A certain portion of the first stone is ground away, but not a great deal, for it falls into the interstices of the pitched stone; they hold each other by the sharp angles; they are kept solid.

Assuming that these [*pointing to a model*] were small lumps of sugar, and that there was an equal space of dough put there, if you put an inch of small lumps of sugar there and an inch of small lumps of sugar on the dough of the same size, and keep rolling a substance over it, which of those would be the soonest to give?—The sugar would sink in the dough.

Would not that tend to unite?—They would make a soft mixture; then you must scrape away that; then you would put on more stone, and the same process would go on, and before you get a solid road you would lose at least twelve or fourteen inches of broken stone.

In your former answer, you said that you thought it would be of advantage if the trustees of the roads were assisted by civil engineering. I wish to know whether road-making ought not to be considered as a branch of the science or art of civil engineering?—I am of opinion that the laying out a good line of road, in some parts of the country, is perhaps as difficult a subject as comes before an engineer; and it is quite impossible for country surveyors and land surveyors, who have not been accustomed to engineering pursuits, to run out a line of road with advantage to the public.

Or short lines for improvements?—I know instances in which lines of road, that are said to be improved, are not so good as the old line of road.

Now, in regard to the construction of a road, is it not necessary a person who undertakes to construct and make a new road, should have that sort of education that makes him acquainted with the science of civil engineering?—It is quite necessary, and that is shown in France to a very great extent; and I believe wherever roads are made here by civil engineers, they bear a very different character to the roads in other parts of the country; and also that there is a saving in the wear of materials on a well-constructed road, and well-laid out road. If a line of road has not rates of acclivity greater than one in forty, there will be 20 per cent. saving over one that rises one in twenty. This is a fact not generally known, but it is quite certain; that is to say, a road that has acclivities of one in twenty will cost 18 or 20 per cent. more than the one that has acclivity of one in forty.

As a matter of course, a civil engineer looks to the appearance his road has, as well as to the fitness to draw carriages; is it not always a rule with them to have it uniform as to breadth and shape?—Yes; there are certain rules which an engineer would always adopt, that is, a certain uniform width, and a certain curvature, a certain height of footway, and a certain width of waste and fences, according to the description of road he was to make.

To acquire that degree of uniformity, is it not necessary to use instruments, and to have that sort of habitual method of managing works that can only be acquired by a regular education?—I conceive so; I do not think a road can well be laid out except by a professional person.

Have you found it the practice to appoint engineers as surveyors of roads?—No; I do not know an instance of it, except on the road between Shrewsbury and Holyhead, and there the effect is very apparent.

What class of persons are they commonly?—Generally farmers; in some instances they have been trades-people.

May not a great deal of what may be considered to be imperfect in the roads of this country, be attributed to the want of having more assistance from the profession?— Yes. I think the fact I have stated, that a saving of 20 per cent. in repairing a road might be made in a very slight alteration of declivity in a road, will prove that principle; and also that there will be a saving when the road has the appearance of uniformity and neatness about it, for the men who put out the stone can see when it is out of shape and where it gets weak, and they instantly repair it. By this means there is not the same jolting, the same degree of resistance, to carriages passing over it; and nothing but a perfect and uniform line of curves and levels will enable the eye of a workman to see where the deficiency takes place.

Generally speaking, the roads are not uniform as to breadth, convexity, or width?—No; there does not appear to be any system in this country on this very important point. Generally speaking, no road, that I am acquainted with, has uniform width and height of footpath and curvature of surface even for half a mile in length.

Would it contribute to the good order and keeping up of a road to pay attention to these points?—It would be a great saving.

A road is easier kept clean and dry?—Yes, and more easily seen when it becomes weak.

Are the workmen more attentive and careful when it is all put in a proper shape?—They would become of a different character; a workman, as soon as he got a uniform and neat road would have a pride in his work, and would keep it in better order and free from ruts, weeds, &c.

Is there not a great deal still wanting in that class of roads for mail-coaches to put them into a proper state; that is, to reduce the hills to such a rate of inclination as may be considered not only convenient but safe, and also in shortening roads?—Yes, there might be a great deal done to improve the roads.

In short, none of the roads are so good as they might be?—I do not think any; and I believe if there was a proper line of section and improvements in the great roads, and if the expenditure was confined to that, instead of a little trifling improvement making by country surveyors, where they run away with a great deal of money, it would be very desirable.

Suppose it was proposed that a line of road in this country should have all the hills brought to an inclination of 1 in 24, and that they should be shortened where it might be done with great advantage, can you form an idea of the average rate of expense per mile necessary to be incurred in making these improvements, and generally in making the roads as perfect as they ought to be made?—I have no doubt that taking a long line of road, from 800*l.* to 1,000*l.* a mile would be sufficient.

What length of line would you take?—From London to York, for example.

Taking a trust with a number of hills and inclinations, would it not amount to 1,500*l.* or 2,000*l.*?—It would depend on the description of the country through which it passed; in some counties it would not, in some of the hilly counties it would come to fully that.

In all hilly counties?—Yes, perhaps it might; no road is perfect unless it has rates of acclivity equal or less than 1 in 40.

Dr. Dionysius Lardner called in, and examined.

You have paid attention to the construction of roads, have you not?—I have.

What, in your opinion, is the proper object and purpose

of a road?—In my opinion the main object of a road connecting two places, is to enable loads to be transported from the one place to the other in the least possible time, and with the least possible expenditure of tractive power.

On what does the tractive power depend?—The tractive power depends upon several qualities in the road; first, upon its levelness; secondly, upon the smoothness of its surface; and, thirdly, upon a quality which I suppose I may call hardness; the absence, in fact, of elasticity.

Is it a matter requiring much science and skill to arrange a road with reference to these objects?—It is quite evident it requires a very unusual combination of scientific and practical knowledge. It is obviously impracticable to make a road which would be theoretically perfect; and therefore there arises an extremely delicate inquiry as to the best possible compromise which can be made between all the inevitable imperfections, the existence of which we are forced to admit. A road, to be theoretically perfect, should be, first, perfectly straight; secondly, perfectly level; thirdly, perfectly smooth; and, fourthly, perfectly hard. If it possessed all these qualities in absolute perfection, the consequence would be it would require no tractive power at all. An impulse given to a load at one end, would carry it to the other by its inertia alone. This is the ideal limit to which it is the business of a road-maker to approximate as nearly as he can, all practical circumstances being considered.

Therefore a road will be more or less perfect in proportion as it approximates to this, all circumstances considered?—Just so. It is obvious it cannot be perfectly straight, and it is obvious it cannot be perfectly level; you might have a perfect level if you chose to make unlimited deviations from perfect straightness; and you might have it perfectly straight, if you chose to encounter the great evil of want of levelness.

Then the degree of straightness would depend upon the degree of hill ?—Just so.

A road ought to be as short as possible, consistently with some regular principle as to hills, ought it not ?—Yes. Now with respect to the acclivities; there is a distinct mechanical character which attaches to acclivities, depending on their steepness. One acclivity is not more injurious than another in the mere ratio in which it is more steep than another. There are some acclivities which afford a certain compensating effect in the descent; there are others that never fully compensate for the power lost in their ascent. There is an acclivity, or an inclination, which we designate in the department of mechanical science that relates to these things by the term of the "angle of repose"; it is the steepest acclivity down which the carriage will not roll of its own accord—down which it will not roll by its own gravity. On more steep acclivities the carriage will roll down without any tractive power; every acclivity under that limit which will require more or less of tractive force downward. Now acclivities, which are less steep than the angle of repose, give a compensation in descending for the excessive tractive force they require in ascending—that is the case with acclivities between the perfect level and the angle of repose; and I take it that that inclination should be the major limit which ought to be imposed to hills, as they are called, upon the first class of turnpike roads; the more they are under that inclination of course the better, but certainly they should never exceed it.

Can you state that acclivity in figures?—That will depend upon several circumstances; it will depend in some measure on the carriage; because a carriage of one structure will roll down a hill, when a carriage of another structure would not. Then it will depend upon the surface of the road; but if we take the very best class of broken-stone road surface, constructed in the best manner so as to be as hard as can be, and a good class of carriage rolling upon it, I suppose, at a rough estimate, one in forty would be the angle of repose. I should advise the great roads not to be more steep than one in forty.

With regard to smoothness; are you of opinion that that should be the only object with respect to roads?—Clearly not; until a comparatively late period a very prevalent, indeed almost universal, error prevailed with respect to roads. All that people considered was what they conceived to be an easy motion to the passengers; that which was easiest to the passengers was concluded to be also the easiest to the horses; or perhaps I should be more correct in saying the horses were not considered at all. People never thought of taking into consideration the mechanical force which was necessary to draw a load along a road. If there were two roads with surfaces equally smooth, (acclivities of an equal steepness, and along which the passengers felt themselves equally comfortable,) those roads were at once assumed to be, to all intents and purposes, equally good; a greater mistake could scarcely be found than that. Suppose a road surface were made of Indian rubber, the surface being as smooth as it can be imagined to be, no road could be worse for traction, the wheels would sink into the surface, and the tractive force would be continually pulling up a hill; it would have the effect of a continual ascent. The surface of the road should be as hard and as unyielding as art can make it; the wheel should not sink; no temporary depression should take place, even though that depression be restored by elasticity after the wheel is removed. By whatever means, this end must be attained.

It is quite essential, although there may be a difference in the means of attaining it; but attained it most certainly ought to be.

Do you speak from experience on this point of elasticity?—Not from experience as an engineer, but only from having devoted a good deal of time to the consideration of this subject, and being perfectly acquainted with the experiments that have been made and the experience we have had upon roads; and I also give that opinion upon general scientific principles.

With reference to the general laws of motion?—Undoubtedly; it is not a point about which any two scientific men can differ; there can be no difference of opinion about it. I mean that that quality is best for the surface of the road which will not permit it to alter its figure under the pressure of the wheel.

Then the degree of hardness will depend upon the degree in which elasticity is absent?—Yes, certainly; a road may have two qualities in that respect; it may yield not being restored, or it may yield with being restored. If it yield without being restored, that will do a double mischief, because, as well as increasing the resistance, the road will be quickly worn out.

There is a question I would wish to ask you with regard to the elasticity: now, glass is a very elastic substance; lead is one which is non-elastic; according to your principle, therefore, if I roll over a plane of glass a glass ball, from those two being elastic it would not go so easily as if I were to roll over it a leaden ball?—Glass is not elastic in that sense. It is elastic with respect to percussion. There are two ways in which you can understand the expression elastic: glass is said to be highly elastic with this meaning; it restores itself to its figure with almost as much force as that with which you alter its figure; but glass is exceedingly hard, and it is difficult to alter the figure of glass.

Upon the same principle, assuming the road yields a little and immediately regains its form, it appears to me to assimilate to the quality of glass?—It is not recovering its form that is of the slightest consequence, because the resistance to the tractive force will be just the same whether it recovers its figure or not; the resistance to the tractive force, strictly speaking, has nothing to do with the elasticity of the road but with its softness; but its being elastic as well as soft does not alter the case in the least. It is difficult to express the thing by those terms, because those are shades of meaning which escape us. It will convey to the Committee more correctly my meaning, if I state that the quality of the road ought to be such that, as the wheels roll over it, it should not suffer any change of its figure.

That is a state of things which is strictly in conformity with the laws of science as relating to moving bodies?—Strictly.

According as I understand you, there is no difference among scientific men as to the necessity of having a road as hard as possible in order to be a perfect road?—Yes, in order to offer the least possible resistance to the tractive force.

Consequently the expense of the draught comes into the calculation in making the road as hard as possible?—Precisely.

Have there not been experiments made upon the railroads with regard to the elasticity?—Yes.

What has been the result?—It has been found that they have considerable elasticity; it is incredible to what an extent this quality exists in them. People would not believe the degree of elasticity there is in an iron railroad; there is a change in passing from one chair to another chair. I have applied an instrument for measuring the tractive force to the waggons upon a railroad, and I have perceived distinctly the passage of the wheels of the waggon over every successive chair of the road.

Have you examined the degree of elasticity of railroads, constructed on other principles, where there has been a continued support to the railroad instead of a support by chairs?—I do not know where that principle has been adopted.

Have you made experiments so as to be able to say what degree of elasticity there is on different railroads?—No; all that I have experimented upon have been constructed upon the same principle, except that some of them are on wooden sleepers and some on stone.

How is it with regard to the elasticity of those two?—I have not compared this; but there is evidently less change from prop to prop on the wooden sleepers.

Is there any doubt that this elasticity requires a greater moving power?—It is a point upon which there can be no doubt.

Would you have the kindness to state to the Committee the difference in the elasticity, when you observed the sleepers were of wood, and when you observed they were of stone?—Inasmuch as the rails are the same, whether placed on wood or on stone, their elasticity cannot be different; but there is this difference, the stone blocks with the chairs upon them form firm unyielding props, and at every yard there is a prop which will not yield under the pressure of the wheel; the rail between the two props consequently is forced to yield, and becomes for the moment a curve, and the wheel rolls down from prop to prop as if it went through the valley of a wave; but with wooden sleepers the pressure which the wheel makes when passing between the props is transmitted to the chairs, and the wooden sleepers yielding to it so that the rail does not become so much curved as with stone blocks.

Then there is less elasticity with the wooden sleepers than with the stone?—Just so; the elasticity is transferred to the wooden sleepers in the one case, instead of the iron rails in the other.

There is more undulation in the one than the other?—Yes; when you pass over certain portions of the Liverpool and Manchester Railway in the carriages you may perceive the undulation.

What, in your opinion, is the proper way of getting rid of the elasticity in roads when you are constructing them? —That is, I think, a question which involves a great deal of practical difficulty; it is quite clear that *cæteris paribus* the thicker the crust of the road is the harder it will be, because a thick crust will not yield as much as a thin one. Then the structure of the road, and nature of the soil, are also to be considered. If a road has a soft subsoil, it has always appeared to me to be that the best method of constructing it is by what I call a *Telfordization*—I mean, in fact, an archway of stone-work under the road; it is a regular piece of archway structure, which abuts on the sides of the road as it were, and upon that the roadwork of broken stone is laid; and the thicker that is, and the more durable and firm the substructure is made, of course the harder will be the road.

And be better with respect to drainage?—And with respect to the drainage it will be better for the duration of the road, of course. It is quite essential to the surface of the road that it should be exposed to the wind and the sun.

Then in laying out a line of road, you would avoid, if possible, going over marshy or elastic ground?—Clearly so; or at least if I did I would take care to press it down so as to destroy its marshy character; as we have done in in the Chatmoss, on the Liverpool and Manchester Railroad.

In flat meadows there is always elasticity, is there not?—Certainly; in fact, what I should say is, the harder the surface is the better will be the road for all purposes. Neither can any thing be more injurious or destructive to a road than the smothering it up with trees and hedges, because every thing that excludes the sun and the air is prejudicial; the water mixes with the dust that is produced by the attrition of the carriages, and that forms a sort of grinding matter that wears away the surface of the road; now if the sun and air have access to the road, the water is quickly evaporated by that means.

You make these observations with reference to the durability of the surface of the road, do you not?—Yes; and also with respect to the tractive force upon it.

Is there more tractive power on a road not exposed to the weather?—There is more tractive power on a road not exposed.

It would appear, from your evidence, that the construction of a road would require a considerable degree of science and practical skill on the part of those who undertake it; it is your opinion that it does so?—I do not know that I could suggest any one problem to be proposed to an engineer, that would require a greater exertion of scientific skill and practical knowledge, than laying out the construction of a road. Unfortunately the original laying out of a road is an employment that is rarely submitted to an engineer; he is generally controlled by circumstances. The early road-makers were almost always obliged to follow our old horse-paths in the country, in a very great degree. To lay out and design a road between two points, the surface of the country should, in the first instance, be accurately ascertained; the engineer should make himself as well acquainted with the undulations and the surface of the country as if he had passed his hand over every foot of it; and, even supposing he has a model of it before him, it becomes an extremely delicate and difficult problem to say what will be the best course to take for a line of road joining two points; he of course must encounter the undulations in such a manner as to adapt his cuttings to his embankments; that is, where he cuts through an eminence he must take care so to arrange the course of the road as that he shall have a hollow to fill up which will just employ the stuff he cuts out of the embankment; then the quality of the crust of the earth he must know, because it is not after he has begun to make his road that he is to discover the practical difficulties which stand in his way. In fact, it requires a considerable knowledge of geology; the stratification and the angles at which different soils will not only stand at the beginning, but the angles at which they will continue to stand, subject to all the actions of the weather.

Then, in point of fact, it does not appear that the advantage of the science of engineering has been applied very extensively to the roads in this country?—Most decidedly not; it never was brought in any degree into play until the problem of railroads was started, when from the nature of the road a greater degree of level and straightness became indispensable; then persons were forced to call in engineering skill, and they found the supply of it totally inadequate to the demand. A board of mushroom engineers have started up for the occasion, and have been forced upon the public. I am quite sure that a vast number of the projects that are now in progress will be very imperfectly carried into effect from the want of that species of mechanical and scientific knowledge which is indispensably necessary.

With reference to general improvements, and also with

reference to new roads in this country, it seems to be your opinion that the assistance of engineers should be more generally called in?—Clearly so; recourse ought to be had to the very first scientific and practical skill of the country; it requires the first civil engineers that can be found.

That applies to putting the existing roads into good repair, independently of making new roads, does it?—Either; but more especially in making new roads.

With regard to the present roads; will they not require considerable improvement in order to render them any thing like what you say a road ought to be?—Undoubtedly they would; even supposing you were not to deviate a single foot from the existing roads, a large portion of them would require re-constructing.

Would that be with reference to hills?—No, I mean with respect to the structure of the road itself; the structure of the crust.

It diminishes the hardness so as to leave the tractive force much greater than it ought to be?—Just so.

Is that a common fault in roads?—That is a very prevalent fault.

Then with regard to hills; would you do any thing with regard to them?—There is a great deal to be done with regard to them.

Still more, perhaps, if it were determined that we were to have roads in the best direction?—I should say, if you were to take any two particular points in England, and were to take the first engineer you could get, and say to him, make the best road between such and such places, without the slightest regard to existing roads, I do not think he would coincide, except by accident, with any of the existing roads; it would be laid down on a totally different principle to what the present roads are laid down upon; the present roads are a series of shifts and expedients which the road-makers have adopted from time to time, without regard to the general scientific principles.

What do you suppose to have been the motives with which the present roads have been laid out in the present lines?—Originally the old horse-paths of the country were followed; and we have kept in the old line of road as long as we can, without great disadvantage. You may observe the most glaring defects take place: you make an occasional deviation with a view to improvement, but return to the original line. There is scarcely an example of one continued line of road, of any length, laid down from end to end on any intelligible principle.

If any plan were attempted by the Legislature for the improvement of the roads in their present state, it is decidedly your opinion that it ought to be conducted under the management of the most experienced civil engineer that this country possesses?—I am quite decidedly of that opinion.

The surveyors generally of roads do not belong to the profession of engineers?—No.

And engineers have not generally been employed in laying out roads?—No; the only cases, and the only species of roads in which great engineering skill has been called into play, are generally the railroads; and even in these there is a great dearth of it.

Are you acquainted with the roads in the Highlands?—Yes.

I would wish to ask you a question about the elasticity of the road. How can it be proved what the elasticity of a road is; take Regent-street for example, at four o'clock in the morning, when there are no persons there; you say it has been ascertained that there is an elasticity in almost every road; I wish to know whether there is any scientific mode of proving that elasticity?—If the elasticity be considerable it will be almost visible; you will almost see the surface yield under the wheels; but if it is less, we have no other means of proving it except by the tractive power.

There has been such a thing as proving the elasticity of a suspension-bridge by a succession of telescopes?—That may be; but it is taking elasticity upon a very extensive scale.

Now take a hard granite road, for instance near London, of which the material is ten inches thick, and which is dry and impenetrable by water, do you conceive that there is any elasticity in a road of that description?—Ten inches is a very considerable thickness for a road-crust of granite; and, if the subsoil be firm and hard, I dare say that that road would be very hard without applying Mr. Telford's method.

Do you mean it would have no elasticity?—It would have very little elasticity.

Does not the tractive power depend, in some measure, upon the friction of the wheels?—The tractive power depends partly upon the carriages; it is different with different carriages; it depends upon the asperities of the roads being encountered by the tire of the wheel; and then, lastly, it depends upon the softness or yielding quality of the road.

Of what stone do you conceive the smoothest surface could be made?—I am scarcely sufficiently acquainted with that subject to answer the question.

Do you think that stone which possesses the greatest smoothness of surface would make the best surface?—I think it would, unless it were deficient in durability.

Is it not apparent that the softest stone would wear into

the smoothest surface sooner than an extremely hard stone?—Yes.

Would you recommend a road to be made of that stone?—I should recommend that the upper stratum of the road should be made of that stone which would wear the smoothest; and that the under stratum should be made of the more durable stone.

You have not probably applied much attention to the effect of different sorts of stones with respect to their smoothness and hardness, and their general effect upon the force of traction?—No, I have not.

I would wish to ask you a question with respect to the vibration upon a road; you stated that you considered the vibration a great repellant to the tractive power; now, in going over a moss, for instance, where there is an evident vibration, do you consider that that vibration is repellant to the tractive power?—I suppose that vibration to be identical, in fact, with what we have been calling elasticity.

Assuming your definition to be correct, that vibration and elasticity are the same, I would wish to ask you whether there would be more vibration or elasticity in a pavement than in a road?—There would be less in a pavement than in a road.

It has been observed in London that a house shakes more from a carriage going along the pavement than it does when the carriage goes along a macadamized road?—That is from the percussion. I should state, perhaps, in speaking of pavement, that good pavement offers the least resistance to the tractive force. I should say that the pavement of the Strand or Fleet-street, when in good order, approaches as near to a railroad as any thing can do.

In a well constructed pavement, such a pavement as

you have alluded to, the more the masonic skill the less the percussion, and the greater the facility, is it not?—Yes.

Therefore the great desideratum in paving is that there should be a very well constructed masonic arrangement?—Yes; but it is surprising, and you would scarcely believe it without a knowledge of the fact, the extent to which that, which has been called vibration, exists even on a railroad. I have experimented on a railroad for the purpose of feeling its surface as well as I could. I have gone in carriages without springs, and it is impossible to convey to you the intolerable sensation I experienced. I thought every limb in my system would be shaken to pieces; and that is the case even over the best railroads. It was produced partly by the joints of the rails on the chairs; there is also, however, some unevenness in the surface of the rail which you can scarcely be aware of; but it may be seen in this way. If you go to a railroad after a heavy shower of rain, and before the wet fully dries, and stand with your face to the sun, so that the rays of the sun striking on the rails will be reflected on the eye, you will see plainly all the unevenness on the surface of the rails, and you will see that they are not inconsiderable.

Have you at all calculated the amount of friction produced by curves on a railroad?—The amount of that friction depends upon the velocity with which the carriage goes, and on the radius of the curve; the friction increases in the same proportion as the square of the speed; that is to say, if you double the speed it will give four times the resistance; and if you treble the speed it will give nine times the resistance, and so on in proportion. It also depends upon the radius of the curve, and it is inversely in proportion with the radius of the curve; the less the radius of the curve the greater will be the resistance.

What is the greatest speed you have ever known to be travelled by carriages on a railroad?—In making experiments I have gone at the rate of 48 miles an hour.

Are there any other observations or suggestions you can make, which you think will be serviceable to the Committee?—None, that I am aware of.

NOTES.

Note A. Page 49.

HOLYHEAD ROAD.

STOWE VALLEY IMPROVEMENT.

*Investigation of the best Plan to be adopted for improving the Road through Stowe Hill Valley. By John Macneill.**

In a great public work of this kind, where a considerable sum of money is to be laid out, it becomes of the greatest importance to ascertain not only what would be the best plan to be adopted, but also to what extent it should be carried, or, in other words, what sum of money should be laid out on the works so as to produce the most advantageous result.

Without altering the entire line of road, as originally proposed by Mr. Telford, which would unquestionably have been the wisest measure, there does not appear to be any means of effecting an improvement of the present line of road, except by embanking across the valley, or lowering the ridges, or by both.

It is evident that each of these plans will admit of different degrees of perfection, according to the sum of money expended on the works; but it is not evident which of these plans is the best, nor does it follow that

* This and the following notes have been furnished by Mr. Macneill.

the same sum of money would produce an equally beneficial improvement, if laid out in raising the valley without lowering the summits, or in lowering the summits without raising the valley. In order to solve this important problem, and to arrive at an accurate result in this and similar investigations, it is necessary to know correctly the expense of horse labour under the varying circumstances of velocity and force of traction on different inclined planes, and also the draught of carriages, and the ratio of the increase of draught in consequence of increase of velocity.

By the experiments lately made on the Holyhead Road by order of the Parliamentary Commissioners, these circumstances have been accurately ascertained from practical experience, which has enabled me to deduce the necessary formulæ from actual practice, without having recourse to theoretical investigations or abstruse calculations.

To go into the detail by which these formulæ were deduced would be in this place unnecessary; it is sufficient to state that correct tables have been calculated from these formulæ, which show the expense of drawing a given weight with a given velocity over every rate of acclivity and declivity, and length of inclined plane.

By means of these tables the expense of drawing a ton weight over any line of road may be determined with great accuracy. Hence all that is necessary in the present investigation is, to calculate by the tables the expense of transporting a ton weight over the existing line of road, and also over the proposed improvements. The difference will be the saving in expense of drawing one ton with the given velocity over the proposed improvement. This, multiplied by the number of tons that pass over the road each day, and by the number of days in the year, will give the annual saving, which, compared with the interest of the money necessary to be expended in

6

making the improvement, will clearly show whether the saving in horse labour is commensurate with the proposed expense. By applying the same criterion to each of the proposed plans, it will at once be made evident which of them should be adopted, as that which would produce the most beneficial result at the smallest expense. By this method, which is new, and founded on correct principles, I have endeavoured to determine the most advantageous method of improving the present line of road across the Stowe Hill Valley from the sixty-fifth milestone to the Crown public-house at Foster's Booth, a distance of two miles.

Plan No. I.

By this plan it is proposed to leave the present road near the sixty-fifth milestone, and to pass at an elevation of twenty-seven feet lower than the present road: from thence it would descend through a natural valley at a rate of inclination of one in thirty to an angle in the Lichborough and Northampton road: from this point it would pass in a straight and horizontal line, at an elevation of fifty feet over the brook, to the junction of the present road; here it would cross the road, and, skirting along the side of the hills at an inclination of one in thirty, running nearly parallel to the present road, and at about fifty yards distance from it, would pass the summit at twenty-seven feet lower level, and join the present road, near the Crown public-house.

For the purpose of ascertaining the comparative merit of this plan, the following calculation, as above described, has been made:—

Pages 439 and 440 contain the calculation of the expense of drawing one ton over the present line of road between the given points, in both directions; which amounts to 82·0647 pence.

Page 441 contains the calculation of expense of drawing one ton over the proposed improved road, as above described, between the same points; which amounts to 76·1724 pence. By this it appears that the saving in horse labour on each ton will be 5·8923 pence; and for 170 tons the daily saving will be 4*l.* 3*s.* 6*d.*, which, at five per cent., is interest for 30,310*l.* 10*s.* The estimate for making this improvement is 23,757*l.* (See page 446.)

The difference between the amount of the estimate and the saving to the public by the proposed improvement is, therefore, 6,553*l.*, which is the actual sum the public would gain by this improvement, supposing the present traffic to continue; if the traffic increased, the saving would be still more.

Plan No. II.

By this plan an embankment is proposed to be raised across the valley, seventy feet high, and 1,313 yards long, the embankment to be formed with earth taken from the most convenient place, without having recourse to cutting from the summits: by this plan part of the present road would be retained.

By a calculation similar to that in the first investigation, and given in pages 441 and 442, it appears that the mean expense of drawing a ton between the sixty-fifth milestone and the Crown Inn at Foster's Booth, would be 79·4584 pence; and as the expense of drawing a ton over the present road between the same points is 82·0647 pence, the saving in expense by this improvement would be 2·6053 pence, and for 170 tons it would be 1*l.* 16*s.* 11*d.*, which is interest for 16,394*l.* The estimate for this improvement, as detailed in p. 447, is 28,890*l.* The difference between this sum and the saving is 12,496*l.*, which is the loss the public would sustain by making this alteration, calculated on the present state of the traffic over the road.

Plan No. III.

By this plan it is proposed to raise the valley fifty feet, to cut twenty-seven feet from the summit at Foster's Booth, and reduce the inclination from 1 in 16, 17, and 18, to 1 in 30, on the east side of the valley; and from 1 in 14 and 15 to 1 in 30, on the west side of the hill.

The embankment would commence near the sixty-fourth milestone, and terminate at the turn of the road leading to Northampton. The mean expense of drawing a ton weight over the road, when improved in this manner, between the sixty-fifth milestone and the Crown Inn at Foster's Booth, as given at p. 443, would be 78·4715 pence: here the saving would be 3·5942 pence on each ton, and for 170 tons it would be 2*l.* 10*s.* 11*d.*, which is interest for 18,483*l.*

The estimate for this work, as detailed in p. 448, is 20,144*l.*; the loss would therefore be 1,661*l.* in making this alteration.

Plan No. IV.

By this plan it is proposed to raise the valley forty feet, to cut twenty-seven feet from the summit at Foster's Booth, and ten feet from the summit at the Angel. The embankment would extend from near the turn of the road to Lichborough to the upper angle of Mr. Drayson's osier plantation, and the lowering of the summit at Foster's Booth would extend into Cold Higham Fields, nearly similar to that in the first plan. All the rest of the road would remain as at present.

If this improvement were made, the mean expense of drawing one ton over it between the sixty-fifth milestone and the Crown Inn at Foster's Booth, would be 79·8450 pence, as shown in p. 444: the saving, therefore, per ton, would be 2·2207 pence, and for 170 tons it would be 377·5190 pence; which would be interest for 11,420*l.*

The estimated expense of this improvement, as detailed in p. 449, is 14,171*l.* The difference is 2,751*l.*; which shows the amount of loss the public would sustain by completing the work here described.

Plan No. V.

By this plan it is proposed to raise the valley forty feet, to lower the summit at Foster's Booth twenty-seven feet, and the summit at the Angel eighteen feet. If this improvement was adopted, the new line would run along the south side of the present road to near the sand pits, where it would cross it obliquely, and entering a small ravine on the opposite side, it would cross the valley in a straight line for the upper angle of the osier plantation. At this point it would again cross to the south side of the present line, and follow the direction described in the first plan.

The saving in expense of drawing a ton over this line of road, when made as above described, would be 5·0200 pence (p. 445), and for 170 tons it would be 475·39 pence, which would be interest for 25,815*l.*

The estimated expense of this improvement, as detailed in p. 450, is 19,607*l.* The difference is 5,208*l.*, which shows the amount of profit the public would acquire by completing the works here described.

Expense of drawing one ton by a stage coach over the present line of road from the sixty-fifth milestone to the Crown Inn at Foster's Booth.

Length.	Rates of Inclination.	Expense.	Length.	Rates of Inclination.	Expense.
300 r.	1 in 43	1·5246	200 r.	1 in 42	1·0200
300 r.	—— 82	1·3872	500 r.	—— 22	3·0415
200 r.	—— 82	·9248	300 r.	—— 20	1·8927
200 r.	—— 40	1·0302	170 r.	—— 43	·8639
100 r.	—— 17	·6774	500 r.	—— 97	2·2735
300 r.	—— 14	2·2539	830 r.	—— 24	4·8978
100 r.	—— 15	·7223	220 r.	—— 24	1·2982
210 r.	—— 15	1·5168	100 r.	—— 18	·6597
130 r.	—— 31	·7112	200 r.	—— 19	1·6754
200 r.	—— 173	·8724	300 r.	—— 23	1·7967
200 f.	—— 197	·7808	300 r.	—— 18	1·9791
500 f.	—— 66	1·6880	300 r.	—— 17	2·0322
350 f.	—— 43	1·0797	200 r.	—— 16	1·3960
500 f.	—— 36	1·5425	200 r.	—— 18	1·3194
200 f.	—— 29	0·6170	400 r.	—— 24	2·3604
300 f.	—— 21	·9255	500 r.	—— 34	2·6730
300 f.	—— 21	·9255	800 r.	—— 64	3·8120
300 f.	—— 20	·9255	200 r.	—— 36	1·0550
200 f.	—— 22	·6170	300 r.	—— 20	1·8927
300 f.	—— 20	·9255	500 r.	—— 34	2·6730
200 f.	—— 20	·6170	250 r.	—— 19	1·6110
300 f.	—— 20	·9255	160 r.	—— 30	·8828
300 f.	—— 20	·9255	250 r.	—— 34	1·3365
300 f.	—— 20	·9255	300 r.	—— 72	2·1333
300 f.	—— 20	·9255	250 f.	—— 119	·9378
280 f.	—— 19	·8638	400 f.	—— 53	1·2340
200 f.	—— 19	·6170	400 f.	—— 37	1·2340
200 f.	—— 18	·6170	200 f.	—— 30	·6170
150 f.	—— 24	·4627	130 f.	—— 19	·4010
200 f.	—— 33	·6170	300 f.	—— 28	
300 f.	—— 56	·9255	200 f.	—— 78	·9255
170 r.	—— 71	·7992	300 f.	—— 78	·7046
190 f.	—— 64	·6348	130.	Horizontal	1·0569
					5360
				Pence	85·7272

Expense of drawing one ton by a stage coach over the present line of road from the Crown Inn at Foster's Booth to the sixty-fifth milestone at Stowe Hill.

Length.	Rates of Inclination.	Expense.	Length.	Rates of Inclination.	Expense.
130	Horizontal ——	0·5360	190 r.	1 in 64 ——	·9053
300 r.	1 in 78 ——	1·3950	170 r.	—— 71 ——	·5860
200 r.	—— 78 ——	·9300	300 r.	—— 56 ——	1·4571
300 r.	—— 28 ——	1·6878	200 r.	—— 23 ——	1·0770
130 r.	—— 19 ——	·8377	150 r.	—— 24 ——	·8851
200 r.	—— 36 ——	1·1036	200 r.	—— 18 ——	1·3194
400 r.	—— 37 ——	2·0968	200 r.	—— 19 ——	1·2888
400 r.	—— 53 ——	1·9592	280 r.	—— 19 ——	1·8043
250 f.	—— 119 ——	1·1172	300 r.	—— 20 ——	1·8927
300 f.	—— 72 ——	1·0377	300 r.	—— 20 ——	1·8927
250 f.	—— 34 ——	·7712	300 r.	—— 20 ——	1·8927
160 f.	—— 30 ——	·4936	300 r.	—— 20 ——	1·8927
250 f.	—— 19 ——	·7712	200 r.	—— 20 ——	1·2618
500 f.	—— 34 ——	1·5425	300 r.	—— 20 ——	1·8927
300 f.	—— 20 ——	·9255	200 r.	—— 22 ——	1·2166
200 f.	—— 36 ——	·6170	300 r.	—— 20 ——	1·8927
800 f.	—— 64 ——	2·6728	300 r.	—— 21 ——	1·8570
500 f.	—— 34 ——	1·5425	200 r.	—— 21 ——	1·8570
400 f.	—— 24 ——	1·2340	200 r.	—— 29 ——	1·1186
200 f.	—— 18 ——	·6170	500 r.	—— 36 ——	2·6375
200 f.	—— 16 ——	·6170	350 r.	—— 43 ——	1·7787
300 f.	—— 17 ——	·9255	500 f.	—— 66 ——	2·3725
300 f.	—— 18 ——	·9255	200 f.	—— 197 ——	·8666
300 f.	—— 23 ——	·9255	200 f.	—— 173 ——	·7744
260 f.	—— 19 ——	·8021	130 f.	—— 31 ——	·4010
100 f.	—— 18 ——	·3085	210 f.	—— 15 ——	·6478
220 f.	—— 24 ——	·6787	100 f.	—— 15 ——	·3085
830 f.	—— 24 ——	2·5605	300 f.	—— 14 ——	·9255
500 f.	—— 97 ——	1·8285	100 f.	—— 17 ——	·3085
170 f.	—— 93 ——	·5244	200 f.	—— 40 ——	·6170
300 f.	—— 20 ——	·9255	200 f.	—— 82 ——	·7116
500 f.	—— 22 ——	·5425	300 f.	—— 83 ——	1·0674
200 f.	—— 42 ——	·6170	300 f.	—— 42 ——	·9255
			Expense of one ton -		78·4023
			Expense of one ton in the contrary direction - - -		85·7272
					2)164·1295
			Mean expense of one ton	Pence	82·0647

Expense of drawing one ton by a stage coach over the road, when improved, as described in Plan No. I.

From the Sixty-fifth Milestone to the Crown Inn at F. B.			*From the Crown Inn at F. B. to the Sixty-fifth Milestone.*		
Length.	Rates of Inclination.	Expense.	Length.	Rates of Inclination.	Expense.
437 r.	1 in 65 —	9·4457	312 f.	1 in 473 —	5·7209
78	Horizontal —	1·4618	1091 f.	—— 30 —	15·2980
758 f.	1 in 30 —	10·6286	1135	Horizontal —	21·2700
1135	Horizontal —	21·2700	758 r.	1 in 30 —	19·0128
1090 r.	1 in 30 —	27·3650	78	Horizontal —	1·4618
312 r.	—— 47 —	5·9708	437 f.	1 in 65 —	6·6716
44 f.	—— 28 —	·6169	28	Horizontal —	·5247
44 f.	—— 78 —	·6606	66 r.	1 in 78 —	1·3949
66 f.	—— 78 —	·9909	44 r.	—— 78 —	·9299
28	Horizontal —	·5247	44 r.	—— 28 —	1·12521
		78·9350			Pence 73·4098
					78·9350
					2)152·3448
			Mean expense of one ton	Pence	76·1724

Expense of drawing one ton by a stage coach over the road, when improved, as described in Plan No. II.

From the Sixty-fifth Milestone to the Crown Inn.			*From the Crown Inn to the Sixty-fifth Milestone.*		
Length.	Rates of Inclination.	Expense.	Length.	Rates of Inclination.	Expense.
300 r.	1 in 43 ——	1·5246	130	Horizontal ——	·5360
300 r.	—— 82 ——	1·3872	300 r.	1 in 78 ——	1·3950
200 r.	—— 82 ——	·9248	200 r.	—— 78 ——	·9300
200 r.	—— 40 ——	1·0300	300 r.	—— 28 ——	1·6878
100 r.	—— 17 ——	·6774	130 r.	—— 19 ——	·8377
100 r.	—— 14 ——	2·2539	200 r.	—— 30 ——	1·1036
100 r.	—— 15 ——	·7223	400 r.	—— 37 ——	2·0968
210 r.	—— 15 ——	1·5168	400 r.	—— 53 ——	1·9592
130 r.	—— 31 ——	·7112	250 r.	—— 119 ——	1·1172
200 r.	—— 173 ——	·8724	300 f.	—— 72 ——	1·0377

TABLE — *continued.*

From the Crown Inn to the Sixty-fifth Milestone.			*From the Sixty-fifth Milestone to the Crown Inn.*		
Length.	Rates of Inclination.	Expense.	Length.	Rates of Inclination.	Expense.
200 f.	1 in 197	·7808	250 f.	1 in 34	·7712
500 f.	—— 66	1·6880	160 f.	—— 30	·4936
350 f.	—— 43	1·0797	250 f.	—— 19	·7712
500 f.	—— 36	1·5425	500 f.	—— 34	1·5425
200 f.	—— 29	·6170	300 f.	—— 20	·9255
300 f.	—— 21	·9255	200 f.	—— 36	·6170
300 f.	—— 21	·9255	800 f.	—— 64	2·6728
300 f.	—— 20	·9255	500 f.	—— 34	1·5425
200 f.	—— 22	·6170	400 f.	—— 24	1·2340
300 f.	—— 20	·9255	200 f.	—— 18	·6170
200 f.	—— 20	·6170	200 f.	—— 16	·6170
5970	Horizontal	24·6143	300 f.	—— 17	·9255
300 r.	—— 23	1·7967	300 f.	—— 18	·9255
300 r.	—— 18	1·9791	300 f.	—— 23	·9255
300 r.	—— 17	2·0322	5970	Horizontal	24·6143
200 r.	—— 16	1·3960	200 r.	—— 20	1·2618
200 r.	—— 18	1·3194	300 r.	—— 20	1·8927
400 r.	—— 24	2·3604	200 r.	—— 22	1·2166
500 r.	—— 34	2·6730	300 r.	—— 20	1·8927
800 r.	—— 64	3·8120	300 r.	—— 21	1·8570
200 r.	—— 36	1·0550	300 r.	—— 21	1·8570
300 r.	—— 20	1·8927	200 r.	—— 29	1·1186
500 r.	—— 34	2·6730	500 r.	—— 36	2·6375
250 r.	—— 19	1·6110	350 r.	—— 43	1·7787
160 r.	—— 30	0·8828	500 r.	—— 66	2·3725
250 r.	—— 34	1·3365	200 r.	—— 197	·8666
300 r.	—— 72	2·1333	200 f.	—— 173	·7744
250 f.	—— 119	·9378	130 f.	—— 31	·4010
400 f.	—— 53	1·2340	210 f.	—— 15	·6478
400 f.	—— 37	1·2340	100 f.	—— 15	·3085
200 f.	—— 30	·6170	300 f.	—— 14	·9255
130 f.	—— 19	·4010	100 f.	—— 17	·3085
300 f.	—— 28	·9255	200 f.	—— 40	·6170
200 f.	—— 78	·7046	200 f.	—— 82	·7116
300 f.	—— 78	1·0569	300 f.	—— 82	1·0674
130	Horizontal	·5360	300 f.	—— 43	·9255
		Pence 83·4388			Pence 75·4780
					83·4388
					2)158·9168
				Mean expense of one ton	Pence 79·4584

Calculation of the expense of drawing one ton from Foster's Booth to Stowe Hill, and from Stowe Hill to Foster's Booth, on the supposition of raising the valley fifty feet, lowering the summit at Foster's Booth twenty-four feet, and reducing the slopes, from 1 in 16, 17, and 18, to 1 in 27 on the south side of the hills, and from 1 in 14 and 15 to 1 in 30 on the north side of the hill, as described in Plan No. III.

Length.	Rates of Inclination.	Expense.	Length.	Rates of Inclination.	Expense.
300 r.	1 in 43 ——	1·5246	130	Horizontal ——	0·5360
1730 r.	—— 30 ——	9·5461	300 r.	1 in 78 ——	1·3950
500 f.	—— 66 ——	1·6880	200 r.	—— 78 ——	·9300
350 f.	—— 43 ——	1·0797	200 r.	—— 28 ——	1·6878
500 f.	—— 36 ——	1·5425	2970	Horizontal ——	12·2453
200 f.	—— 29 ——	·6170	300 f.	1 in 20 ——	·9255
300 f.	—— 21 ——	·9255	200 f.	—— 36 ——	·6170
300 f.	—— 21 ——	·9255	300 f.	—— 64 ——	2·6728
300 f.	—— 20 ——	·9255	500 f.	—— 34 ——	1·5425
200 f.	—— 22 ——	·6170	400 f.	—— 24 ——	1·2340
300 f.	—— 20 ——	·9255	2760 f.	—— 30 ——	8·5146
200 f.	—— 20 ——	·6170	3630	Horizontal ——	14·9664
300 f.	—— 20 ——	·9255	600 r.	1 in 20 ——	3·7854
600 f.	—— 20 ——	1·8510	300 r.	—— 20 ——	1·8927
3630	Horizontal ——	14·9664	200 r.	—— 20 ——	1·2618
2760 r.	1 in 30 ——	15·2297	300 r.	—— 20 ——	1·8927
400 r.	—— 24 ——	2·3604	200 r.	—— 22 ——	1·2166
500 r.	—— 34 ——	2·6730	300 r.	—— 20 ——	1·8927
800 r.	—— 64 ——	3·8120	300 r.	—— 21 ——	1·8570
200 r.	—— 36 ——	1·0550	300 r.	—— 21 ——	1·8570
300 r.	—— 20 ——	1·8927	200 r.	—— 29 ——	1·1186
2970	Horizontal ——	12·2453	500 r.	—— 36 ——	2·6375
200 f.	1 in 28 ——	·6170	350 r.	—— 43 ——	1·7787
200 f.	—— 78 ——	·7046	500 r.	—— 66 ——	2·3725
300 f.	—— 78 ——	1·0569	1730 f.	—— 30 ——	5·3370
130	Horizontal ——	·5360	300 f.	—— 43 ——	·9255
From Stowe Hill to F. B.	Pence	80·8594	From F. B. to Stowe Hill	Pence	76·0836
					80·8594
					2)156·9430
			Mean expense of one ton	Pence	78·4715

Calculation of the expense of drawing one ton by a stage coach from the Crown Inn at Foster's Booth to the sixty-fifth milestone, and from the sixty-fifth milestone to the Crown Inn at Foster's Booth, on the supposition of raising the valley forty feet, cutting twenty-seven feet from the summit at Foster's Booth, and reducing the inclination on Lovell's Hill from 1 in 14 and 15 to 1 in 30, as described in Plan No. IV.

Length.	Rates of Inclination.	Expense.	Length.	Rates of Inclination.	Expense.
300 r.	1 in 43	1·5246	130	Horizontal	·5360
1730 r.	—— 30	9·5461	300 r.	1 in 78	1·3950
500 f.	—— 66	1·6880	200 r.	—— 78	·9300
350 f.	—— 43	1·0797	200 r.	—— 28	1·6878
500 f.	—— 36	1·5425	2970	Horizontal	12·2453
200 f.	—— 29	·6170	300 f.	1 in 20	·9255
300 f.	—— 21	·9255	200 f.	—— 36	·6170
300 f.	—— 21	·9255	800 f.	—— 64	2·6728
300 f.	—— 20	·9255	500 f.	—— 34	1·5425
200 f.	—— 22	·6170	400 f-	—— 24	1·2340
300 f.	—— 20	·9255	200 f.	—— 18	·6170
200 f.	—— 20	·6170	200 f.	—— 16	·6170
300 f.	—— 20	·9255	300 f.	—— 17	·9255
300 f.	—— 20	·9255	300 f.	—— 18	·9255
400 f.	—— 20	1·2340	300 f.	—— 23	·9255
3960	Horizontal	16·3671	260 f.	—— 19	·8021
650 r.	1 in 24	3·8357	100 f.	—— 18	·3085
100 r.	—— 18	·6597	650 f.	—— 24	2·0052
260 r.	—— 19	1·6754	3960	Horizontal	16·3271
300 r.	—— 23	1·7967	400 r.	1 in 20	2·5236
300 r.	—— 18	1·9791	300 r.	—— 20	1·8927
300 r.	—— 17	2·0322	300 r.	—— 20	1·8927
200 r.	—— 16	1·3960	200 r.	—— 20	1·2618
200 r.	—— 18	1·3194	300 r.	—— 20	1·8927
400 r.	—— 24	2·3604	200 r.	—— 22	1·2166
500 r.	—— 34	2·6730	300 r.	—— 20	1·8927
800 r.	—— 64	8·8120	300 r.	—— 21	1·8570
200 r.	—— 36	1·0550	300 r.	—— 21	1·8570
300 r.	—— 20	1·8927	200 r.	—— 29	1·1186
2970	Horizontal	12·2453	500 r.	—— 36	2·6375
200 f.	1 in 28	·6170	350 r.	—— 43	1·7787
200 f.	—— 78	·7046	500 r.	—— 66	2·3725
300 f.	—— 78	1·0569	1730 f.	—— 30	5·3370
130	Horizontal	·5360	300 f.	—— 43	·9255
From Stowe Hill to F. B.		Pence 81·9941	From F. B. to Stowe Hill		Pence 77·6959
					81·9941
			Mean expense of one ton.		2)159·6900
					Pence 79·8450

Calculation of the expense of drawing one ton by a stage coach over the road, when improved, as described in Plan No. V.

From Stowe Hill to F. B.			*From F. B. to Stowe Hill.*		
Length.	Rates of Inclination.	Expense.	Length.	Rates of Inclination.	Expense.
437	1 in 65 r. ——	9·4457	286	1 in 85 r. ——	5·9879
758	—— 30 f. ——	10·6286	231	—— 130 r. ——	4·6431
1,135	Horizontal ——	21·2710	1,130	—— 30 f. ——	15·8448
1,130	1 in 30 r. ——	28·3437	1,135	Horizontal ——	21·2710
231	—— 138 f. ——	3·9951	758	1 in 30 r. ——	19·0129
286	—— 85 f. ——	4·6555	437	—— 65 f. ——	6·6716
44	—— 28 f. ——	·6169	28	Horizontal ——	·5247
44	—— 78 f. ——	·6606	66	1 in 78 r. ——	1·3949
66	—— 78 f. ——	·9909	44	—— 78 r. ——	·9299
28	Horizontal ——	·5247	44	—— 28 r. ——	1·1252
		Pence 81·1327			Pence 77·4060
					81·1327
					2)154·0914
				Mean expense of one ton - - -	77·0457

Plan No. I.

Estimate of the expense of making a new line of road between Stowe Hill and Foster's Booth, by which the valley would be raised fifty feet, and the greatest rate of inclination in any part would not exceed 1 in 30.

	£	s.	d.
To 266,524 cubic yards of earth-work, at 1*s*. - - - -	13,326	0	0
To 24,238 cubic yards of earth-work, at 6*d*. - - - -	605	0	0
To 20 acres of land for site of new road and embankment, at 100*l*. per acre -	2,000	0	0
To 1,135 yards of fences over the embankment, at 10*s*. - - - - -	567	10	0
To 2,935 yards of post and rail fences, at 6*s*. - - - -	888	10	0
To 4,070 yards of road-making, at 15*s*. -	3,052	10	0
To culverts, 8 feet wide, under embankments - - - -	616	0	0
To cross-drains - - - - -	50	0	0
To field-gates - - -	100	0	0
To toll-house and gates - - - -	400	0	0
	21,597	10	0
Contingencies, to cover unforeseen expenses	2,159	10	0
	£23,757	0	0

Plan No. II.

Estimate of the expense of raising the valley seventy feet.

	£	s.	d.
To 668,512 cubic yards of earth-work in the embankment, at 7*d.* - -	19,498	5	4
To 25 acres of land to be purchased to supply earth for the embankment, at 100*l.* - - - -	2,500	0	0
To 14 acres of land required for the site of the embankment, at 100*l.* - -	1,400	0	0
To 1,313 yards of fences over the embankment, at 10*s.* - - - -	656	10	0
To 1,313 yards of road-making, at 15*s.* -	984	15	0
To culvert, 8 feet wide, under embankment - - - -	824	0	0
To toll-house and gates - - -	400	0	0
	26,263	10	4
Contingencies - - -	2,626	9	8
	£28,890	0	0

Plan No. III.

Estimate of the expense of raising the valley fifty feet, lowering the summit at Foster's Booth twenty-seven feet, and reducing the slopes from 1 in 16, 17, and 18, to 1 in 30, on the east side of the valley; and from 1 in 14 and 15 to 1 in 30, on the west side of the valley.

	£	s.	d.
To 104,671 cubic yards of earth-work, at 1*s.* - - - -	5,233	11	0
To 207,781 cubic yards of earth-work, at 7*d.* - - - -	6,060	5	7
To 31,784 cubic yards of earth, at 6*d.* -	794	12	0
To 13½ acres of land, to be purchased for the new lines of road and embankments, at 100*l.* per acre - - - -	1,350	0	0
Ten acres of land to be purchased to supply earth for the embankments, at 100*l.* - - - -	1,000	0	0
To 682 yards of fences over the embankments, at 10*s.* - - -	341	0	0
To 1,766 yards of post and rail fence, at 6*s.*	529	16	0
To 2,450 yards of road-making, at 15*s.* -	1,837	10	0
To culvert, 8 feet wide, under the embankment - - - - -	616	0	0
To cross-drains - - -	50	0	0
Field-gates - - - - -	100	0	0
Toll-house and gates - - -	400	0	0
	18,312	14	7
Contingencies - - - - -	1,831	5	5
	£20,144	0	0

Plan No. IV.

Estimate of the expense of raising the valley forty feet, cutting twenty-seven feet from the summit at Foster's Booth, and reducing the inclinations from 1 in 14 and 15 to 1 in 30 on the west side of Lovell's Hill.

	£	s.	d.
To 104,671 cubic yards of earth-work, at 1*s*. - - - - -	5,233	11	0
To 12,440 cubic yards from back cutting, at 9*d*. - - - -	466	0	0
To 63,931 cubic yards from back cutting, at 7*d*. . - - -	1,864	13	1
To 2,611 cubic yards of earth from back cutting, at 6*d*. - - -	65	5	6
To three acres of land to be purchased for supplying earth for embankment, at 100*l*.	300	0	0
To twelve acres of land to be purchased for the site of embankments, and for the cutting at Foster's Booth and Lovell's Hill, at 100*l*. - - - -	1,200	0	0
To 616 yards of fences over the embankment, at 10*s*. - - - -	308	0	0
To 1,870 yards of post and rail fence, at 6*s*.	561	0	0
To 2,486 yards of road-making, at 15*s*. -	1,864	10	0
To eight feet culvert under embankment	520	0	0
To cross-drains - - -	50	0	0
To field-gates, &c. - - -	50	0	0
To toll-house and gates - - -	400	0	0
	12,882	19	7
Contingencies - - - - -	1,288	0	5
	£14,171	0	0

Plan No. V.

Estimate for the expense of raising the valley forty feet, cutting twenty-seven feet from the summit at Foster's Booth, and fourteen from the summit at Stowe Hill.

	£	s.	d.
To 198,200 yards of earth-work in embankments taken from the cuttings, at 1*s.*	9,910	0	0
To 23½ acres of land for cuttings and embankments, at 100*l.* - - -	2,350	0	0
To 682 yards of fences on the embankments, at 10*s.* - - -	341	0	0
To 3,374 yards of post and rail fences, at 6*s.* - - - - -	1,014	4	0
To 4,055 yards of road materials, at 15*s.*	3,041	5	0
To 8 feet culvert under embankment -	520	0	0
To cross-drains - - -	100	0	0
Field-gates - - - - -	150	0	0
Toll-house and gates - - -	400	0	0
	17,824	9	0
Contingencies - - - - -	1,782	11	0
	£19,607	0	0

General Abstract.

No. of Plan.	Estimate. £	s.	d.	Saving in Horse Labour. £	s.	d.	Saving to the Public. £	s.	d.	Loss to the Public. £	s.	d.
1.	23,757	10	0	30,310	10	0	6,553	0	0			
2.	28,890	0	0	16,394	0	0	- - - -			12,496	0	0
3.	20,144	0	0	18,483	0	0	- - - -			1,661	0	0
4.	14,171	0	0	11,420	0	0	- - - -			2,751	0	0
5.	19,607	0	0	25,815	0	0	5,208	0	0			

NOTE B. Page 67.

THE resistance produced by collision is seldom a constant retarding force; loose stones, or hard substances, are sometimes met with, and will give a sudden check to the horses, according to the height of the obstacle: the momentum thus destroyed is often very considerable.

The power required to draw a wheel over a stone or other obstruction may be thus determined:—Suppose A B D to represent a carriage wheel fifty-two inches in diameter, the axis 2·5 inches in diameter, the weight of the wheel 200 lbs., and the load on the axle 300 lbs. Let a stone or other obstacle four inches high be represented as at S; the power necessary to be applied to the axle to draw the wheel over the stone is thus found:—Suppose P the power which is just sufficient to keep the wheel balanced, or in equilibrio, when acting from the centre C in the direction C P. The force acting against this power is gravity, and is equal to the weight of the wheel and load on the axle, acting from the centre C in the direction C B. These forces act together against the point D in the direction C D. Gravity acts in the direction C B with the energy or length of lever D B, and the power acts in the direction C P, with the leverage B C; and the equation of equilibrium will be $W \ DB \ P \times CB$. In this equation, C B the radius of the wheel diminished by the height of the obstacle, and B D equals $\sqrt{DC^2 - BC^2}$; hence the power $P = \frac{W \times \sqrt{DC^2 - BC^2}}{DC - DS}$: in the present example,

$W = 500$; $DC = \frac{52}{2} = 26$; $BC = DC - DS = 26 - 4 = 22$; and $\sqrt{DC^2 - BC^2} = 13 \cdot 85$; the power, therefore, which is necessary to keep the whole in equilibrium, or resting off the ground, supported at the point D, will be equal to $\frac{500 \times 13 \cdot 85}{22} = 314 \cdot 3$ lbs.

The pressure at the point D is equal to the joint action

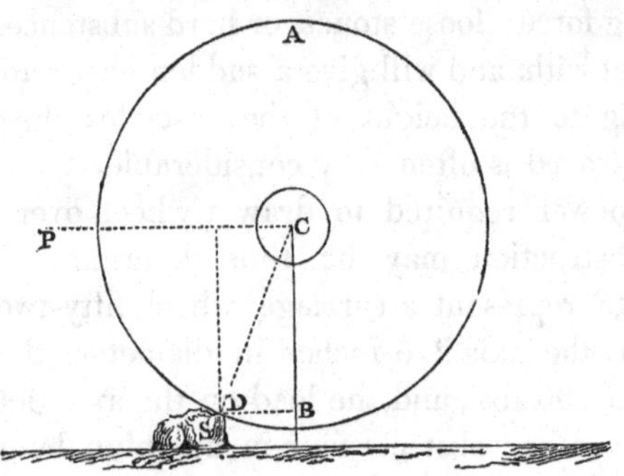

of the power and weight, as before stated; this in the present instance is represented by the radius of the wheel: hence $CB : CD :: 500 : \frac{CD \times 500}{CB} = \frac{26 \times 500}{22} = 591$ lbs. nearly. The injury which a road sustains by this pressure acting on a small point, and in an oblique direction, is very great: but it is not alone in this that the road suffers: the force with which the wheel strikes the surface, in its descent from the top of the stone is considerable, and would soon wear a hole in the hardest road. But it must be observed, that a carriage mounted on proper springs will be drawn over an obstacle of this kind with much less power than if the carriage had no springs; for the springs allow the wheels to mount over the obstacle without raising the body of the carriage, and its road along with it, to the same height: upon this principle alone it is that carriages mounted on proper springs are easier moved than those without springs; and, for the same reason, springs are more necessary on rough and uneven roads than on smooth ones; and in proportion to the roughness of the roads should the springs be free and elastic; and it is to the improvement in the roads, of late years, that the rigid elliptic springs on carriages have been substituted for the C springs formerly in use; for, when ruts were very numerous, and the surface of the roads rough and uneven, the springs that are now used would have been of very little use, as their vertical motion is so limited, besides having no lateral play.

By enlarging the diameter of the wheel, the power required to draw it over an obstacle will be diminished; and, should the weight of the wheel remain the same, the power will decrease nearly as the diameter of the wheel increases; we have seen that, when the wheel was fifty-two inches in diameter (which is the general size of the front wheels of waggons), it required about 314 lbs. to keep the wheel in equilibrium when resting on a point four inches above the general surface. Suppose, now, that the hind wheel (the diameter of which is sixty-four inches) is to be pulled over the same obstacle, the power will be found to be only 305 lbs., although the weight is increased 70 lbs. by enlarging the wheel;

$$\text{for } P = \frac{W \times \sqrt{DC^2 - BC^2}}{DC - DS} = \frac{570 \times 15 \cdot 5}{28} = 305.$$

In this investigation we have supposed the power to act directly from the centre, and parallel to the horizon, which supposition is sufficient for practical purposes; but, if great accuracy be required, it will be necessary to allow for the thickness of the axletree, and to diminish the length of the lever by which the power and weight act; for suppose the power to balance the weight, the point of contact of the axle with the nave will be at half a right angle from the vertical line, or the arch ef will be 45°,

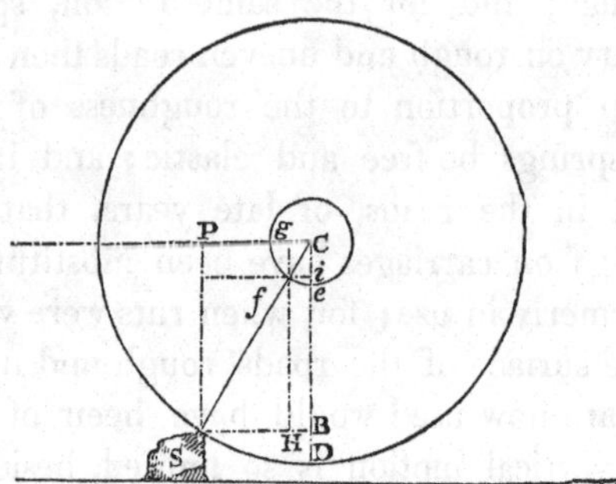

and f H will represent the force and direction of gravity, and S H that of the power; but f H is equal to $CB - Ci$, and S H is equal to $SB - fi$, and because fiC is a right angle, $fi = Ci$. If the radius of the axle CF be 1·5 inches, the sine fi, or Ci, of 45°, will be 1·06 inches: hence $SH = SB - 1{\cdot}06 = 13{\cdot}83 - 1{\cdot}06 = 12{\cdot}79$, and $fH = CB - fg = 22 - 1{\cdot}06 = 20{\cdot}94$; and the power $P = \frac{500 \times 12{\cdot}79}{20{\cdot}94} = 305{\cdot}4$ lbs.; which is 10 lbs. less than that before found, without allowing for the diameter of the axle.

If the power be applied to the axle in a direction not parallel to the horizon, but inclining upwards, as represented in the following figure, the resistance will be diminished, or a less power will be required; for the leverage by which the power acts is increased, while the leverage by which gravity acts is decreased, until the line of draught forms a right angle with the line drawn from the centre of the axle to the point of contact, in which case the power is a minimum. On the contrary, if the direction of the line of draught be inclined downwards from the horizontal, the leverage by which it acts will be diminished, and consequently the power must be increased as the direction of the line of draught falls below the horizontal.

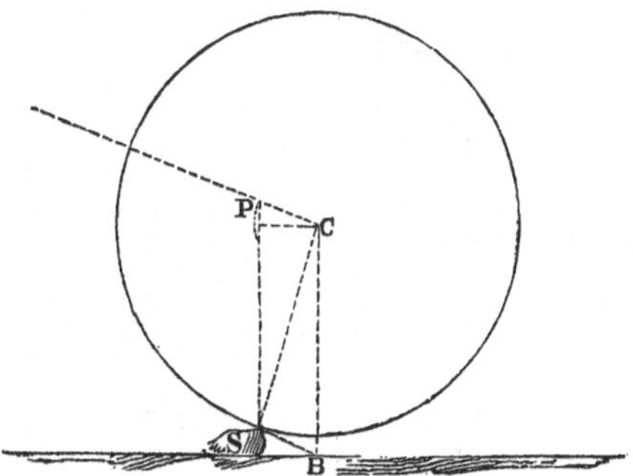

Formerly it was the practice to fasten on the streaks of iron, or shoeing, on the wheels of carriages with nails, the heads of which projected at least half an inch beyond the surface of the wheel. These heads formed a succession of obstacles over which the wheel had to mount; and, besides being extremely injurious to the roads, were a serious obstacle to the effective work of horses: the iron shoeing is now generally put on in hoops, and fastened by rivets, the heads of which are countersunk, and therefore form no impediment to the rolling of the wheel.

When the surface is indented, or furrowed into small cavities, such as a pebble pavement or badly dressed stone (into which the wheel falls, producing a shock, the noise of which is well known), the resistance which is produced arises from a different cause. For the momentum or velocity in the horizontal direction is partly destroyed by the descent of the wheels into the hollows and the blow or collision against the opposite side of the cavity.

The resistance produced by such a surface has also been investigated by M. Gerstner. His reasoning may be briefly stated as follows:—

Suppose B E D one of the cavities formed by two contiguous stones.

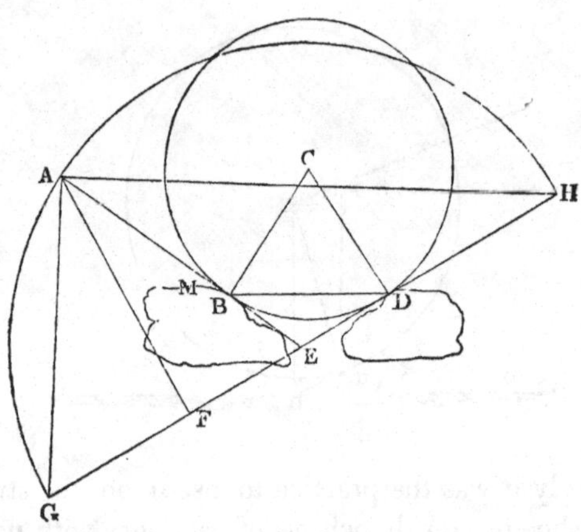

Let the tangents B E and D E be drawn to the circumference of the wheel at the points of contact B and D, and suppose the velocity to be represented by A E=H E in magnitude and direction.

From the point E as a centre, and with the radius A E, describe the semicircle G A H, and let fall the perpendicular A F. The velocity A E may be resolved into two others, A F and F E: of these two, one, A F, is destroyed by the shock, and the other, F E, remains acting in the direction E D; consequently, the loss of velocity is equal to A E−E F=E G−E F=G F; and this loss must be compensated by an increase in the force of traction.

To avoid a complicated calculation, suppose the force of traction, K‴, to be a constant accelerating force; Q, the weight of the carriage; and $2\,g\,t$, the velocity which gravity would generate in the weight Q at the end of the time t: we shall then have

$K''' = \frac{Q \cdot FG}{2\,g\,t}$. But FG : AG :: AG : 2 AE; from which

$FG = \frac{AG^2}{2AE}$; and from the similar triangles AEG, DCB; $AG : AE :: DB : BC$; from which $AG = \frac{AE \times DB}{BC}$; but $t = \frac{MN}{u}$; r representing the velocity with which the space MN is traversed in the time t. By making these substitutions, the former equation becomes $K''' = \frac{Q v^2}{4 g MN} \left(\frac{DB}{BC}\right)^2$.

From this expression the following conclusions may be drawn: —

1. That the resistance arising from a surface of this description is proportional to the load.

2. That the draught or force of traction is proportional to the square of the velocity; and consequently pebble or rough pavements are more adapted for heavy loads, with a slow velocity, than for light carriages with quick velocities.

3. That the draught increases in the inverse ratio of MN; that is, as the distance between the paving-stones diminishes, or as the stones are narrow, the cavities remaining the same.

4. That the draught increases in the ratio of the width of the cavity to the radius of the wheel.

When the stones made use of for paving are of a good shape, well dressed, and sufficiently large, and laid on a firm and substantial foundation, they form the most perfect road surface for general purposes. The cavity between the stones should not exceed half an inch in width, by which means carriage wheels would pass over them without the least shock or resistance, and consequently without producing the noise often complained of in towns, at the same time that the surface would be sufficiently rough to prevent the horses from slipping.

Note C. Page 72.

The next resistance, friction, which we shall consider, is that which arises from the wheels being forced over obstacles which break down under their weight, or when they are drawn through mud or other soft substances, or when the material of which the road is composed (such as gravel or small stones) is put on a soft or yielding substratum, in layers so thin that the weight of the wheel can make an impression on it.

Let A B C represent a carriage wheel resting on the

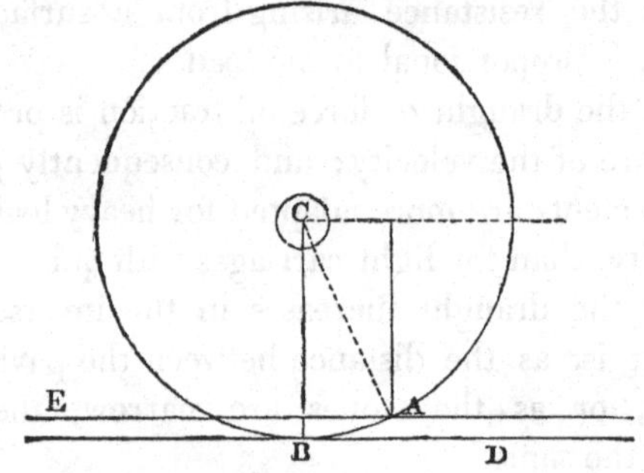

horizontal road B D, the surface of which is hard and solid, but covered with mud, sand, or gravel, to the height of the line A E: if it be very soft, the wheel, as it rolls along, will press through it as if it was water, and rest on the hard and firm surface B D. If it be of a more tenacious nature, as some clays, or composed of sand or gravel which the wheels will only compress, without displacing it, the wheel will not go to the hard surface, but approach it in proportion to the weight on the axle or wheel and the compressibility of the material over which it passes. A heavy wheel will sink deeper than a light one into a soft road, if both wheels be of the same dimensions. At the point A, where there is no weight, the surface is undisturbed; and at the point B, the material composing the road is compressed and sunk as much as it can be by such a weight: all the intermediate part between A and B is pressed by a less weight, decreasing from B towards A, and is proportionally compressed or lowered. The resistance which is opposed to the wheel evidently arises from its action upon that portion of sand or mud contained between A and B; and the power necessary to overcome this will depend upon the length of lever at which it acts, or the depth to which the wheel sinks, and the stiffness or incompressibility of the substance which covers the road. Hence it is impossible to say or calculate the power or draught necessary to draw a carriage over a road so circumstanced, without experiments being made to ascertain the compressibility of each substance, and the consequent effect on the draught of carriages with wheels of different construction, and different loads. It is, however, within the power of mathematical investigation to furnish formulæ by which the law of increase in the power necessary to overcome such resistances is known, and by combining these with experiments the power necessary to draw a carriage over any line of road may be determined.

If the resistance arises from the wheel sinking into a soft stratum, instead of through an accumulation of mud or dust, until it rests on a firm surface, the investigation will be similar: the only difference is, that in one case the wheel can only sink a limited depth; for, arriving at the hard surface of the road, it can penetrate no farther. The leverage at which the power acts will remain constant, if the weight be sufficient to press the wheel through the soft covering to the solid surface. The resistance will depend upon the nature of the material through which it rolls; but, if there be no solid or hard substratum under the outer crust, there will be no limit to the depth to which the wheel will sink. Thus, when a cart is drawn through a ploughed field, it is well known that the wheels will penetrate to a depth proportionate to the load, and the labour of the horses will be increased accordingly.

This effect is nearly the same as that which takes place when a carriage is drawn over a weak gravelly road, and is evidently more injurious to the horses employed in draught than when they work on a solid and firm road, although it be covered with an inch or two of mud.

M. Gerstner has investigated this subject, and given formulæ for the resistance arising from a wheel passing over a soft stratum of different degrees of compressibility. These formulæ are,

$$K''=\frac{fq}{A}=\frac{q^2}{A\,h\,m};\ \text{when } m=o;$$

$$K''=\frac{3fq}{4A}=\tfrac{3}{4}q\sqrt[3]{\frac{3q}{2A^2bW}};\ \text{when } m=1;$$

$$K''=\frac{5fq}{8A}=\tfrac{5}{8}q\sqrt[5]{\frac{15q}{8A^3bm}};\ \text{when } m=2:$$

in which K'' = the resistance;

f = half the chord of the segment of the wheel in the ground;

A = radius of the wheel;

q = weight of load;

W = resistance of the soil when compressed by the wheel to the depth of an inch or any other unit;

m = an indeterminate number, expressing a power of the depth to which the wheel penetrates, proportional to the resistance of the soil at that depth, and which is to be determined by experiment.

From these formulæ it is evident that the resistance is caused by the wheels sinking into the ground; and therefore it will be better, under such circumstances, to divide a heavy load between two or more carriages than to carry a heavy load on one carriage; and also that the resistance will be diminished by increasing the width of the tiers of the wheels.

Note D. Page 44.

When the road is not horizontal, the force of gravity is a great impediment to the draught of carriages, and limits considerably the effect which would otherwise be produced by a horse in drawing a load.

If it were not for the hills that are usually met with on turnpike roads, one horse would do as much work as four; for it is well known that the force of draught must be increased in proportion to the steepness of hills: the quantity of that increase is thus determined:—

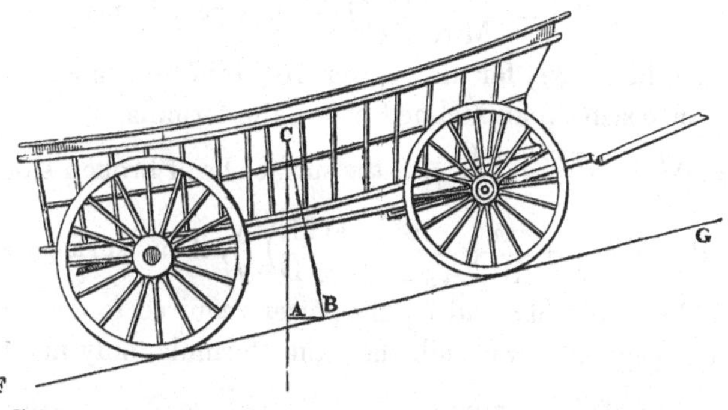

Suppose a waggon resting on an inclined plane, F G; and let C be the centre of gravity of the waggon and load. Draw the line C B perpendicular to the surface of the hill, and C A perpendicular to the horizon; let this last line represent the force of gravity, or weight of the carriage and load. This force is equivalent to two others represented by the lines A B and C B in magnitude and direction. The force represented by C B is the pressure of the carriage on the surface of the road, and that represented by A B is the force, independent of friction, which acts against the carriage going up hill, or tends to force it down hill.

Now this force may be found as follows:—The triangle A B C is similar to the triangle A K F; for the angle F A K = the angle C A B = the angle E G F; and the angles A B C and A K F are each right angles; therefore A C : A B :: F A : A K; but F A : A K as the length of the plane is to its height; that is, A C : A B :: l : h; and as the line A C represents the weight of the carriage, or W, we have W : A B :: l : h; or, $AB = \frac{Wh}{l}$ = the force represented by the line A B. The power required to draw a carriage on a horizontal may be represented by the formula $P = W\left(\frac{rm}{R} + \frac{3f}{4R}\right)$ for a broken-stone road, and $P = W\left(\left(\frac{rm}{R}\right) + \frac{v^2}{4gMN}\left(\frac{DB}{BC}\right)^2\right)$ for a paved one.*

We then have, for the power required to draw a waggon or coach up an inclined plane, the formula,
$P = W\left(\frac{h}{l} + \frac{rm}{R} + \frac{\frac{3}{4}f}{R}\right)$, if the surface be of broken stone;
or $P = W\left(\frac{h}{l} + \frac{rm}{R} + \frac{v^2}{4gMN}\left(\frac{DB}{CD}\right)^2\right)$, if the surface be a pebble pavement; and for the power required to draw the same waggon down hill, the same formula, only making the sign of $\frac{h}{l}$ negative.

In these formulæ, W = the weight of the waggon wheels and load: for although it might at first sight appear that we should make use of the weight on the axle, or that represented by the line C B, to calculate the resistance, yet it is not so; for the pressure on the axles will be equal to the joint action of the weight on the axles and the moving power, and this will be the force represented by the line A C or W, so that no correction of the weight is necessary.

The resistances arising from part of the weight being

* See Mémoire sur les Grandes Routes, &c, de M. F. de Gerstner.

thrown from the front axles to the hind ones, in consequence of the inclination of the traces, and the line of draught not passing through the centre of gravity of the carriage, may be omitted in a general investigation; also the correction that should be applied to the resistances where the carriage is on an inclined plane; because it is evident that there is less weight on the surface than if the carriage stood on level ground, and also from the hind wheels bearing a greater pressure than the front ones, in consequence of the line of gravity falling nearer to the hind wheels; as the difference that will take place in the draught, in consequence of these, will be inconsiderable in general practice, and, should extreme accuracy be required in any particular case, it will be easy to make the necessary calculations.

The following experiments were made with the waggon the axles and wheels of which had been previously made use of for the experiments on friction.

1. Half a ton of stone was put in the waggon, as nearly as possible in the centre between each axletree; the waggon was then drawn over a timber platform, perfectly horizontal, by weights suspended from a line: to effect this, it required 50¾ lbs.

2. A ton of stone was placed in the waggon, half a ton over each axletree, and the power required to draw the waggon over the same surface was 70 lbs.

3. A ton and a half of stone was placed in the waggon, and distributed equally over each axletree; the weight or power required to draw the waggon was then found to be 90 lbs.

The resistances arising from the friction of the axletrees in the above experiments were then calculated for each wheel from the formula before given, and the total resistance arising from the axles, thus determined, was subtracted from the draught or power found by experiment as requisite to draw the waggon; the difference gave the resistance of the surface caused by the penetration of the wheels into the timber surface.

The results of these experiments are given in the following Table:—

Weight of Waggon and Load in Pounds.	Power required to draw the Waggon.	Resistance of the Axles.		Resistance of the Surface.
2,240	31·0	13·0 10·6	23·6	7·4
2,800	52·0	16·2 13·3	29·5	22·5
3,360	70·0	19·5 15·9	35·4	34·6
3,920	91·0	2·7 8·6	41·3	49·7

By a considerable number of experiments with the same waggon, on roads of different kinds, the draught was found to agree very nearly with the results calculated from the empyrical formula,

$$P=\frac{W+w}{93}+\frac{w}{40}+v;$$

in which W = the weight of the waggon; w, the load; c, a constant number, which will depend on the surface over which the waggon is drawn; and v, the velocity in feet per second. By putting $v=3\cdot 7$, which was the velocity used in the foregoing experiments, the constant number for a timber surface was determined, and found to be equal to 2.

For other surfaces, the value of c may be taken as follows:—

On a paved road - - - - - - 2
On a well-made broken stone road in a dry clean state - - - - - - 5
On a well-made broken stone road covered with dust 8
On a well-made broken stone road wet and muddy - 10
On a gravel or flint road in a dry clean state - - 13
On a gravel or flint road in a wet muddy state - 32

On an inclined plane the above formula becomes $P=\frac{W+w}{93}+\frac{w}{40}+v+\frac{h}{l}.\ \frac{W+w}{1}$; for a common stage waggon, and $P=\frac{W+w}{100}\times\frac{w}{40}\times\ c\frac{h}{l}.\ \frac{W+w}{1}$ for a stage coach.

THE END.

London:
Printed by A. Spottiswoode,
New-Street-Square.

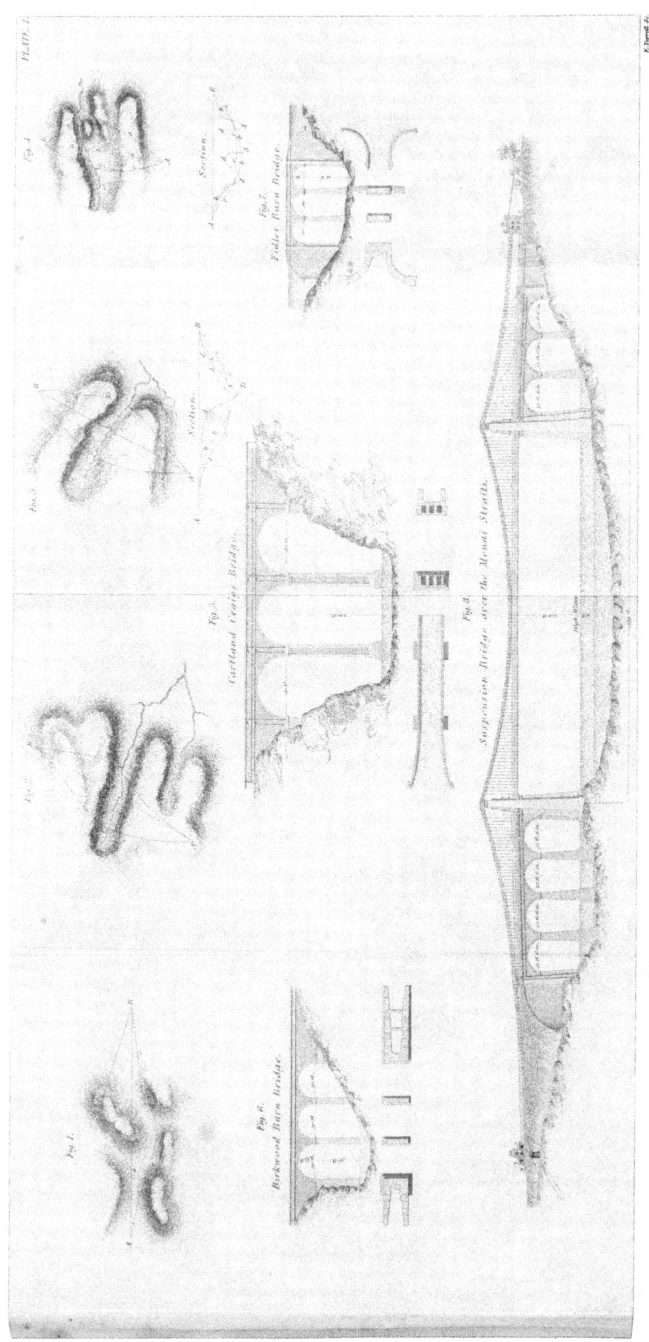

ne material originally positioned here is too large for reproduction in this issue. A PDF can be downloaded from the web address given on page iv this book, by clicking on 'Resources Available'.

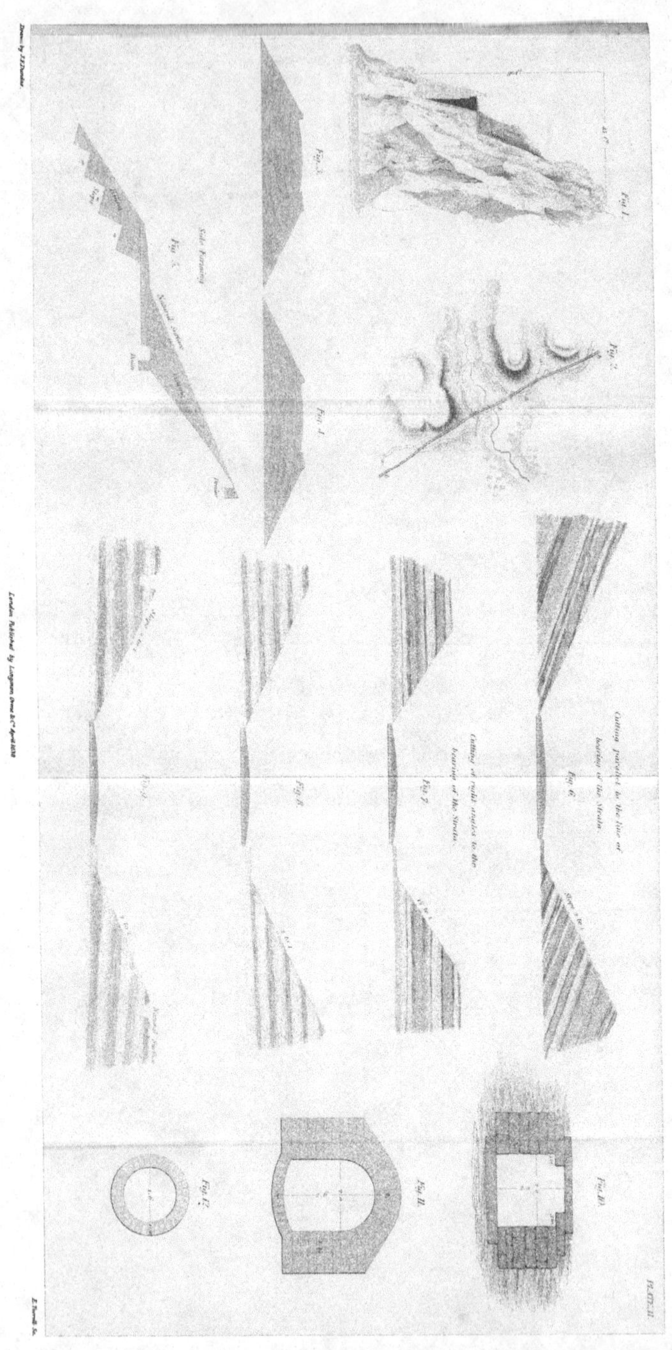

The material originally positioned here is too large for reproduction in th
reissue. A PDF can be downloaded from the web address given on page i
of this book, by clicking on 'Resources Available'.

e material originally positioned here is too large for reproduction in this issue. A PDF can be downloaded from the web address given on page iv this book, by clicking on 'Resources Available'.

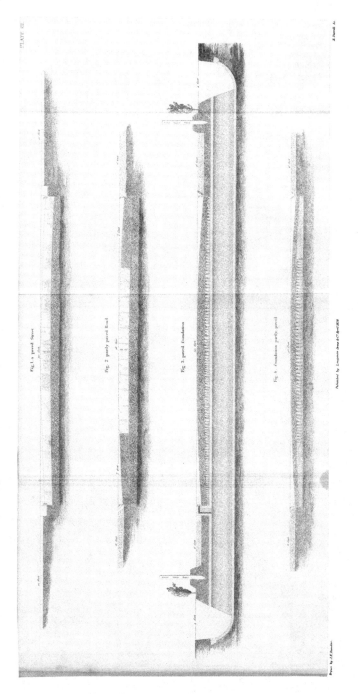

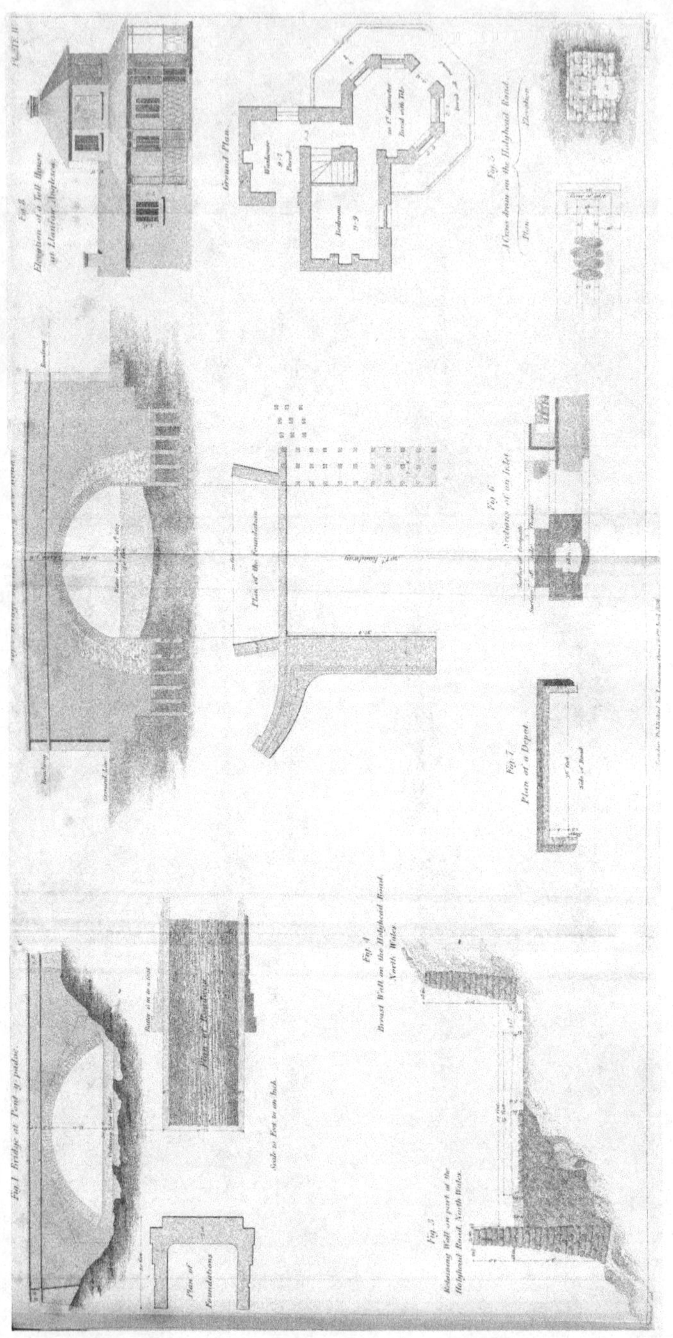

The material originally positioned here is too large for reproduction in th
reissue. A PDF can be downloaded from the web address given on page i
of this book, by clicking on 'Resources Available'.

e material originally positioned here is too large for reproduction in this issue. A PDF can be downloaded from the web address given on page iv this book, by clicking on 'Resources Available'.

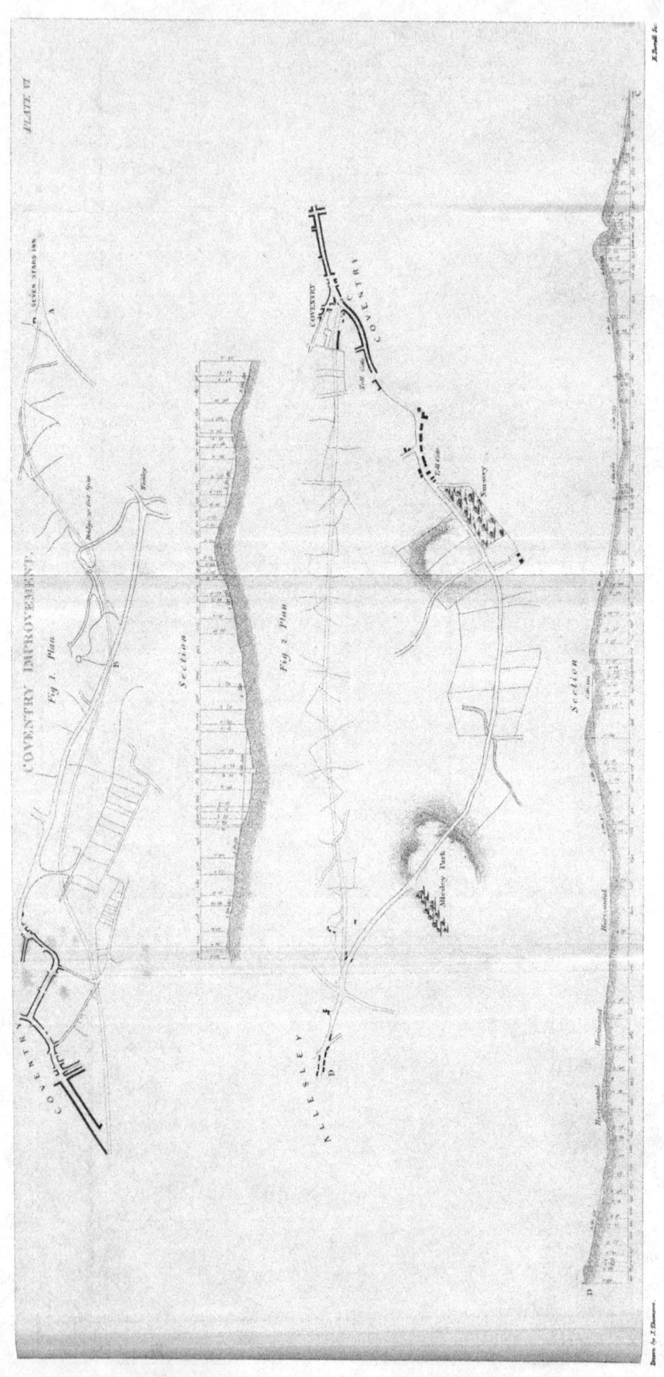

The material originally positioned here is too large for reproduction in th
reissue. A PDF can be downloaded from the web address given on page i
of this book, by clicking on 'Resources Available'.

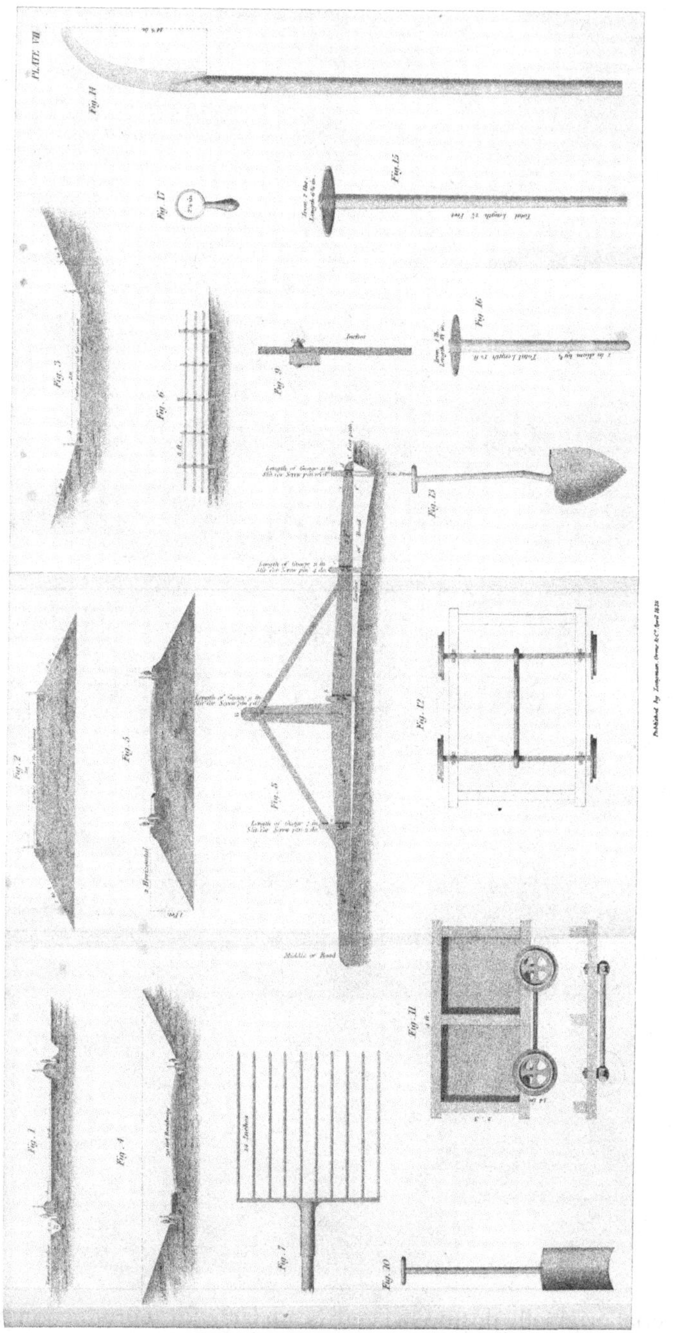

e material originally positioned here is too large for reproduction in this
issue. A PDF can be downloaded from the web address given on page iv
this book, by clicking on 'Resources Available'.

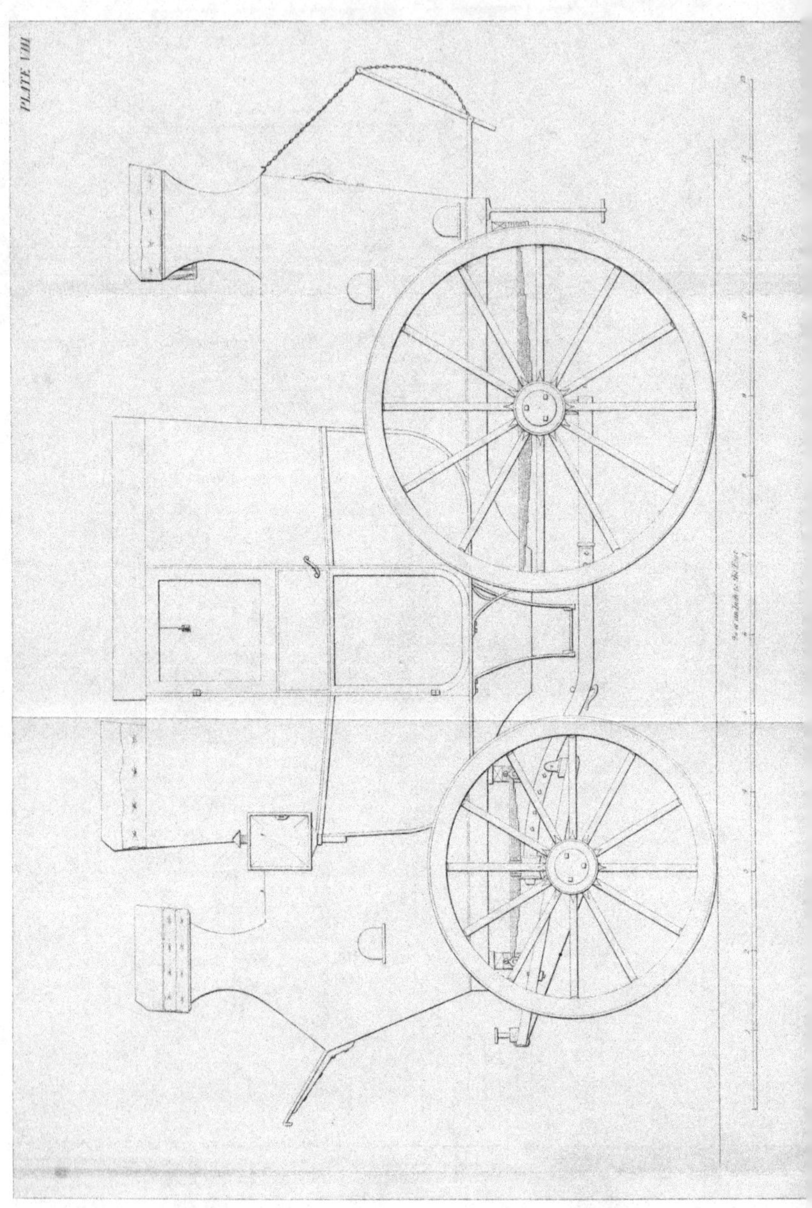

The material originally positioned here is too large for reproduction in th
reissue. A PDF can be downloaded from the web address given on page i
of this book, by clicking on 'Resources Available'.

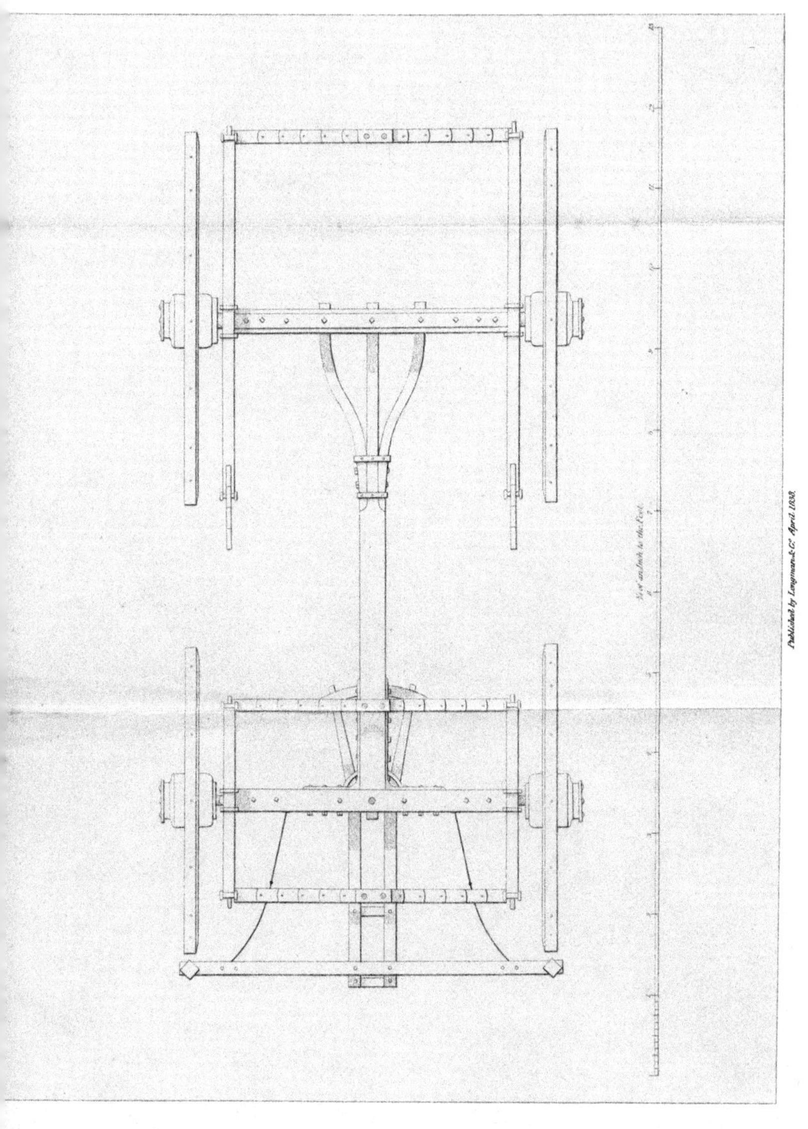
Published by Longman & Cº April 1838.

e material originally positioned here is too large for reproduction in this ssue. A PDF can be downloaded from the web address given on page iv this book, by clicking on 'Resources Available'.

Printed in the United States
By Bookmasters